NMR of Newly Accessible Nuclei
Volume 1

CHEMICAL AND BIOCHEMICAL APPLICATIONS

NMR of Newly Accessible Nuclei
Volume 1

CHEMICAL AND BIOCHEMICAL APPLICATIONS

Edited by

PIERRE LASZLO

Institut de Chimie
Université de Liège
Liège, Belgium

1983

ACADEMIC PRESS

A Subsidiary of Harcourt Brace Jovanovich, Publishers
New York   London
Paris   San Diego   San Francisco   São Paulo   Sydney   Tokyo   Toronto

ACADEMIC PRESS, INC.
111 Fifth Avenue, New York, New York 10003

United Kingdom Edition published by
ACADEMIC PRESS, INC. (LONDON) LTD.
24/28 Oval Road, London NW1 7DX

Library of Congress Cataloging in Publication Data

Main entry under title:

NMR of newly accessible nuclei.

  Includes bibliographies and indexes.
  Contents: v. 1. Chemical and biochemical applications
-- v. 2. Chemically and biochemically important elements.
  1. Nuclear magnetic resonance spectroscopy. I. Laszlo,
Pierre. II. Title: N.M.R. of newly accessible nuclei.
QD96.N8N58 1983   543'.0877   83-4619
ISBN 0-12-437101-9 (v. 1)

Contents

QD96
N8 N58
1983
v.1

CHEM

A. GENERAL FEATURES

1. Multinuclear Instrumentation and Observation
C. Brevard

2. Techniques That Can Enhance Sensitivity, Improve Resolution, Correlate NMR Spectral Parameters, and Lead to Structural Information
Philip H. Bolton

3. Computational Considerations
Charles L. Dumoulin and George C. Levy

4. Factors Contributing to the Observed Chemical Shifts of Heavy Nuclei
G. A. Webb

5. Quadrupolar and Other Types of Relaxation
R. Garth Kidd

B. SELECTED FEATURES

6. Cation Solvation
Robert G. Bryant

10. Antibiotic Ionophores

Hadassa Degani

Contributors

Numbers in parentheses indicate the pages on which the authors' contributions begin.

Philip H. Bolton (21), Department of Chemistry, Wesleyan University, Middletown, Connecticut 06457

C. Brevard (3), Bruker Spectrospin S. A., 5-67160 Wissembourg, France

Robert G. Bryant (135), Department of Chemistry, University of Minnesota, Minneapolis, Minnesota 55455

Hadassa Degani (249), Isotope Department, The Weizmann Institute of Science, Rehovot, Israel

Torbjörn Drakenberg (157), Physical Chemistry 2, University of Lund, S-220 07 Lund, Sweden

Charles L. Dumoulin (53), Department of Chemistry, Syracuse University, Syracuse, New York 13210

Sture Forsén (157), Physical Chemistry 2, University of Lund, S-220 07 Lund, Sweden

R. Garth Kidd (103), Department of Chemistry, University of Western Ontario, London, Ontario N6A 5B7, Canada

George C. Levy (53), Department of Chemistry, Syracuse University, Syracuse, New York 13210

Björn Lindman (193, 233), Physical Chemistry 1, University of Lund, S-220 07, Lund 7, Sweden

Hans J. Vogel (157), Physical Chemistry 2, University of Lund, S-220 07 Lund, Sweden

G. A. Webb (79), Department of Chemistry, University of Surrey, Guildford, Surrey GU2 5XH, United Kingdom

General Preface

Books must follow sciences, and not sciences books.
Francis Bacon

In heeding this recommendation from the author of "The Advancement of Learning," we capitalize upon the spectacular progress of nuclear magnetic resonance (NMR) in recent years. Not only have NMR methods reached beyond chemistry and biochemistry to fertilize biomedical research, but more recently even radiologists and clinicians have started using NMR imaging techniques to diagnose and to monitor treatment. Chemistry and biochemistry have also benefited from recent advances, which have been numerous, impressive, and very useful. The introduction of Fourier transform and pulse excitation techniques has had all the features of a genuine mutation. As a consequence we have witnessed a steady and impressive gain in spectrometer sensitivity, together with a delightful improvement in ease of operation: one can now "dial" nearly any nucleus and communicate with it in a manner of minutes, in contrast to the changes of probe configuration that required hours in the 1960s and 1970s.

Beyond such quantitative changes, there are also qualitative changes, especially the availability of literally dozens of sophisticated and extremely useful pulse sequences (whose acronyms would require separate booklets to define and explain their workings). As a consequence, NMR spectroscopy has acquired new dimensions. Two-dimensional NMR is now (almost) routine, and some ancient prohibitions have fallen by the wayside. High-resolution NMR of solids, multiple quantum NMR, and many other very sophisticated modes of spectral acquisition have been perfected through the pioneering work of the likes of Richard Ernst, Ray Freeman, and John Waugh, to whom I wish to pay homage here.

This work, in two volumes, focuses on the "newer" nuclei, those that only a few years ago were not readily accessible with continuous wave excitation. For this reason, they were then nicknamed "exotic nuclei" or "other nuclei," to set them apart from the more familiar nuclei: ^{1}H, ^{13}C, ^{15}N, ^{19}F, and ^{31}P. Even though this distinction between the familiar and the exotic is already outdated and on the wane, we have set it as our goal to provide a state-of-the-art report on the newer nuclei. In so doing, we avoid duplicating the excellent existing monographs for the familiar nuclei.

Another pitfall we wished to avoid was an attempt at presenting an all-embracing work on the chemical applications of NMR. It was possible for the Emsley, Feeney, and Sutcliffe volumes to achieve such comprehensiveness in the 1960s, but today this task would require thousands of pages. Hence, we have elected to emphasize methods of study ("Methods are the habits of the mind and the savings of the memory"—Rivarol). There are two ways to present these methods: the problem-oriented approach and the technique-oriented approach. Accordingly, our first volume is devoted to general principles of magnetic resonance relative to the new nuclei and to special applications selected for their importance and timeliness. The second volume is a systematic survey, consisting of concise, albeit rather comprehensive treatments, of some of the most important nuclei and families of nuclei in the periodic table. Such a two-pronged treatment should help provide in-depth coverage. This is a novel approach, commending itself, we feel, by its simplicity and appropriateness. In this manner the reader will be able to find information more quickly and effectively.

A central feature of the work is worthy of special comment. The book lays special emphasis on the exploitation of relaxation processes, both as a new dimension of NMR that has come to the fore in the 1960s and 1970s and as a source of all-important parameters for studying the thermodynamics and kinetics of binding. Thus the "three dimensions" of a spectral line frequency, scalar couplings, and relaxation rates are treated not as discrete entities but as inseparable elements in a single informational continuum.

We hope that these two volumes will serve as handbooks to be found in every NMR laboratory. But their natural readership, besides the NMR experts who we hope will find these volumes a useful summary, will consist of chemists and biochemists wishing to get initiated into the magnetic resonance methods at the graduate level. Indeed, Volume 1 might serve as the text for a one-semester introductory course in NMR aimed at first-year graduate students, with Volume 2 as the companion reference book. Lecturers in the chemistry and biochemistry departments of schools of medicine and pharmacy will find the books appropriate for such courses as self-contained texts. The two-volume format of presentation will also be an aid in the ''self-teaching'' of the new NMR.

Preface to Volume 1

In this first volume of "NMR of Newly Accessible Nuclei" we document how multinuclear NMR is being used in practice. The easy access to very many nuclear resonances has allowed organometallic chemists during the period 1970–1980 to perform structural and kinetic studies similar to those that were already current practice in organic chemistry during the previous decade using predominantly proton NMR. Inorganic and analytical chemists have used NMR of the newer nuclei to gain fundamentally new knowledge, not only about details of coordination chemistry, but also about ionic solvation, as described in the chapter by R. G. Bryant. In this volume, applications are indeed centered upon ions. Biochemists and biophysicists have learned, mostly through the pioneering work of the Swedish school, to use the resonance of halides and of alkali metal nuclei for binding studies (chapters by H. J. Vogel, T. Drakenberg, and S. Forsén, and by B. Lindman). The competition between binding and the less specific atmospheric condensation characterizes interactions of ions with polyelectrolytes and micelles, which the Swedish school has also done so much to elucidate and which we are also fortunate to have presented here by one of its leaders (B. Lindman). Membranes, of central concern in the life sciences, can be traversed by ions, if the ions are transported by carrier molecules. The story of the contribution of NMR to the study of antibiotic ionophores is told by H. Degani.

Besides presenting such chemical and biochemical applications, we provide the reader with a "how to" manual, so that he or she can set about doing work of the same caliber. In the first part of the present volume we have emphasized operation of a modern NMR spectrometer under its two aspects: the instrumental (C. Brevard) and the computational (C. L. Dumoulin and G. C. Levy). A chapter is devoted to some of the most important pulse sequences, for example, INEPT (P. H. Bolton). The other two chapters review the various factors determining the position of the observed absorption (G. A. Webb) and those responsible for the various relaxation processes (R. G. Kidd).

The topics presented in this volume represent a personal choice. This has the possible advantage of providing some internal coherence and the certain disadvantage of bias which may appear to privilege certain areas unduly and to neglect others. I take full blame for the selection, hoping nevertheless that the reader will find the menu of this *table d'hôte* appealing, with enough variety that he or she may want to sample more than the one chapter that happens to be of particular interest at the moment.

In closing, I should like to acknowledge gratefully the initial suggestion from Professor Peter Stang, of the University of Utah, with whom I wrote a book on organic spectroscopy 13 years ago, that I should undertake this endeavor. I am also extremely grateful to my co-authors, who have enthusiastically accepted their assignments, despite their involvement with many other activities (which, I was pleased to learn during my editing of these two volumes, besides biochemical and chemical research of the first rank, range from carpentry and cello design to running a cultural center, sailing, and canoeing!). Finally, I wish to thank my secretary, Madame Nicole Dumont-Troisfontaines, for her patient and calm efficiency in collecting and collating the manuscripts.

Contents of Volume 2

A

General Features

1 Multinuclear Instrumentation and Observation

C. Brevard

Bruker Spectrospin S.A.
Wissembourg, France

I. Introduction

The potential of NMR spectrometers to perform multinuclear observations has opened up a promising new field of chemistry. An increasing number of results are being obtained via NMR studies on "other" isotopes, but often when starting such experiments, there are two samples to deal with, the chemical substance *and* the spectrometer. It is the aim of this chapter to delineate the general organization of a multinuclear spectrometer and to stress the difficulties and drawbacks to be expected in multinuclear research.

II. The Multinuclear NMR Spectrometer

When starting NMR experiments on nuclei other than ^1H, ^{19}F, ^{13}C, or ^{31}P, for which chemical shift scales and clear-cut operating conditions are well known, the experimentalist faces some unusual situations such as large chemical shifts (18,000 ppm for Co), large coupling constant values (140,000 Hz for $^1J_{199_{Hg}199_{Hg}}$, Granger, 1982), and very broad lines (quadrupolar isotopes). These conditions require full flexibility of the spectrometer to allow for an adequate recording strategy. Continuous-wave or stochastic excitation has been surpassed by the pulse excitation method because of the ease of the latter technique, and recent electronic developments have made it possible to cope with practically any situation.

This chapter will deal with pulse Fourier transform (FT) spectrometers. Figure 1 is a diagram of such a multinuclear system.

A. The Pulse FT Method

Since its first description (Ernst and Anderson, 1966), the pulse FT method has been widely adopted, but it is of interest to stress some of the limitations of the technique in relation to the previously mentioned multinuclear recording requirements.

Apart from the well-known folding-back phenomenon, which can lead to misleading interpretations, the pulse method introduces a parameter of prime importance, namely, the 90° pulse width value $\tau^X_{90°}$ for a given isotope X and its associated frequency domain.

According to the mathematical expression for the FT of a box function, if ν_0 represents the carrier frequency, the power spectrum of a repetitive pulse train of pulse duration τ and pulse interval T is represented by an envelope of n harmonic frequencies, $\nu_i = \nu_0 \pm n/T$, the amplitudes of which obey a law

$$A_i = A_0(\sin \pi\tau \ \Delta\nu)/\pi\tau \ \Delta\nu \tag{1}$$

with $|\Delta\nu| = |\nu_0 - \nu_i| = n/T$ in hertz.

Although proton observation generally fulfills the condition of constant power distribution across the entire spectrum (short τ values, small $\Delta\nu$ spectral width values), this situation no longer holds in a search for an unknown X resonance over a very large spectral width. Figure 2 depicts in a dramatic manner such a power spectrum for single-channel detection and $\tau = 30 \ \mu s$ (90°) and 10 μs (30°).

The systematic use of quadrature detection, which allows the carrier frequency to be set in the middle of the explored spectral width, doubles

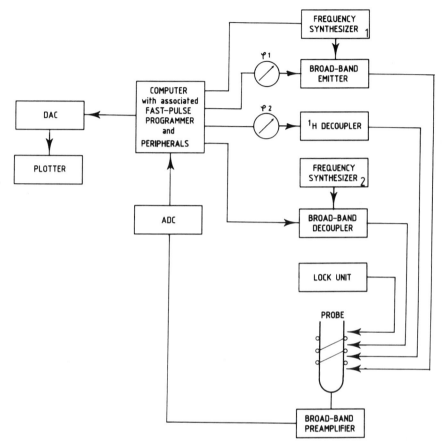

Fig. 1 Diagram of a multinuclear NMR spectrometer.

the usable portion of the power distribution (Fig. 3). Nevertheless, care should be taken to avoid any blank portion in the spectrum, namely, ν_i values near the τ^{-1}-hertz region. A good compromise is to choose small τ values together with quadrature detection for such a wide-sweep observation, to profit from the corresponding power distribution.

On the other hand, apart from the narrow frequency range covered, long pulse width values introduce very serious phase problems when employing elaborate multipulse sequences such as INEPT (Morris and Freeman, 1979) and INADEQUATE (Bax *et al.*, 1980a, 1980b), which will find increasing use in multinuclear NMR (see Chapter 2). A careful probe design will help to eliminate, or at least attenuate, the burden on the experimentalist.

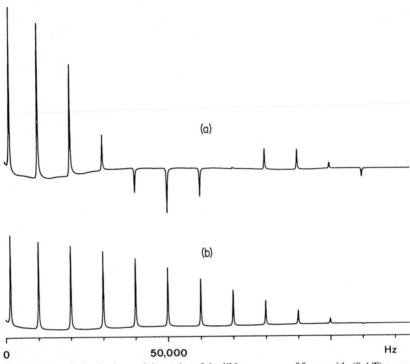

Fig. 2 The variation in the peak intensity of the ^{14}N resonance of formamide (9.4 T) across a 125,000-Hz sweep width according to the pulse width value. (a) 30-μs (90°) pulse; (b) 10-μs pulse. The carrier offset has been sequentially shifted by 10,000-Hz steps after every 100 scans in a single-detection mode. The $(\sin x)/x$ dependence is clearly visible in (a), whereas spectrum (b) benefits from a more constant power distribution because of a smaller pulse width.

Another important parameter in multinuclear observation is the relaxation time T_1 of the isotope under study. The most extreme situations occur when two isotopes of the same element (e.g., ^{187}Os and ^{189}Os) have widely differing T_1 values (100 s and 10 μs, respectively). An evaluation of the T_{1X} value is mandatory before performing any experiment. Section III will give some suggestions for deciding on the right isotope.

B. The Computer and the Associated Pulse Programmer

This accessory is one of the most important parts of a multinuclear FT spectrometer. Although the computer word length is not as critical as for proton observation, because of less severe dynamic range problems, flexi-

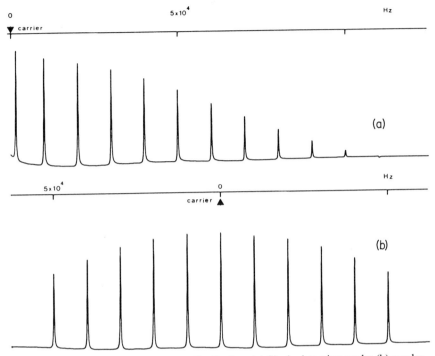

Fig. 3 Comparison of the pulse power distribution. (a) Single-detection mode; (b) quadrature detection mode. Parameters and recording conditions as in Fig. 2b.

bility is important. In fact, all commercially available FT spectrometers possess a computer with a dedicated fast-pulse programmer and fast analog-to-digital converter (ADC) which allow the following:

(a)　decoupler, emitter, and receiver phase switching,
(b)　multilevel decoupler settings,
(c)　simultaneous emission of emitter and decoupler pulses with inter-pulse delays ranging from 1 μs to 1 h,
(d)　large sampling rate up to 150 kHz,
(e)　intelligent management of peripherals such as disk units, magnetic tapes, variable-temperature unit or laser beams.

The introduction of direct memory access technology or parallel processors interfaced with the central processing unit has improved capabilities to such a point that a true time-shared operation of spectrometers is now possible with respect to acquisition, processing, plotting, computing, and spectrum identification.

The use of a fully computer-controlled broad-band emitter and pream-

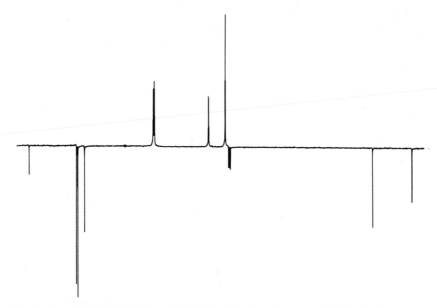

Fig. 4 The 400-MHz ¹H (up) and 100-MHz ¹³C (down) spectra of 10% ethylbenzene in CDCl₃ recorded via a doubly tuned probe and computer-switchable ¹H and ¹³C preamplifiers. The unique FID contains both ¹³C and ¹H information.

plifiers and double-tuned probe heads is illustrated in Fig. 4 where the ¹H and ¹³C spectra were obtained without manual readjustments. This automation achievement is useful, for example, when studying biological samples that require alternate ³¹P and ¹H acquisitions within the short sample lifetime in the NMR tube.

C. The Broad-Band Emitter

To detect the resonances of most magnetic isotopes, the spectrometer must be capable of generating a wide range of frequencies (typically 8–400 MHz at 9.39 T). The chosen frequency is then amplified and pulsed via the broad-band emitter. The nominal pulse angle α_X is related to the emitter rf field B_1 by

$$\alpha_X = \gamma_X B_1 \tau / 2\pi \qquad (2)$$

The need for a short $\tau^X{}_{90°}$ (Section II,A) imposes a B_1 field value as large as possible, as one expects following Eq. (2) an *increase* in $\tau^X{}_{90°}$ with *decreasing* γ_X values. Modern broad-band emitters deliver about 100–150

W with a progressive increase in the power output for low frequencies to provide in practice a constant $\tau^X_{90°}$ in the low-frequency range.

Selective excitation pulse sequences such as the DANTE (Bodenhausen *et al.*, 1976; Morris and Freeman, 1978) sequence can be implemented by simply inserting either a manual or a computer-controlled attenuator at the broad-band emitter preamplification stage.

D. *The Decoupler*

Apart from its normal decoupling purpose, the proton decoupler is finding increasing use as part of multipulse, multiphase INEPT-like sequences (Morris and Freeman, 1979; Bendall *et al.*, 1981) which require pulse capabilities for the decoupler itself. Pulsing the decoupler means a computer-controlled decoupling power setting with associated digital-to-analog converter (DAC), and especially very good pulse shaping and decoupler rf phase stability to avoid catastrophic sensitivity losses during the experiments. Figure 5 shows a straightforward calibration of the decoupler $\tau^{1H}_{90°}$ pulse using the X part of an AX doublet following the sequence

$$90°_X-d-\alpha_H-\text{acquisition} \tag{3}$$

with $d = 1/2\ J_{XA}$.

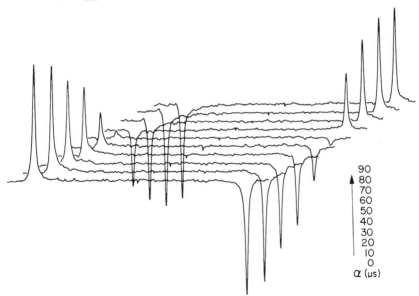

$$
\begin{array}{c}
90 \\
80 \\
70 \\
60 \\
50 \\
40 \\
30 \\
20 \\
10 \\
0 \\
\alpha\ (\mu s)
\end{array}
$$

Fig. 5 Determination of the decoupler $\tau^{1H}_{90°}$ pulse width value following sequence (3). Sample: $CHCl_3$; X = ^{13}C; $d = 0.0238$ s. When α is varied, the null value gives $\tau^{1H}_{90°} = 50\ \mu s$.

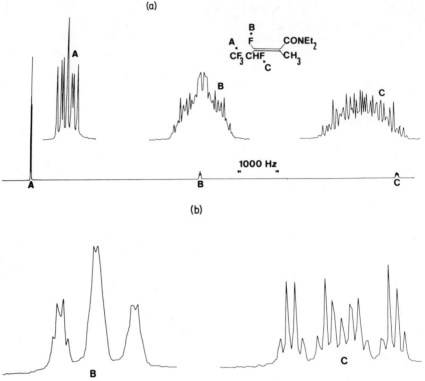

Fig. 6 The 84.6-MHz ^{19}F spectrum of the compound shown in the inset. (a) Normal spectrum; (b) with ^{19}F homodecoupling of trifluoromethyl group A. Fluorine atoms B and C show clearly long-range F—H and F—F coupling. Broad-band proton decoupling eventually collapses B and C patterns into two doublets ($J_{F_BF_C}$). (Sample courtesy of Pr. Wakselmans, CNRS, Thiais.)

On the other hand, {^{1}H} and/or {Y} heterodecoupling or {X}—X homodecoupling will certainly prove very useful in the near future, as {^{31}P} INEPT (Brevard and Schimpf, 1982), {^{103}Rh}—^{13}C (Heaton *et al.*, 1980), {^{109}Ag}—^{1}H (Van Stein *et al.*, 1982, or {^{29}Si}—^{29}Si (Harris *et al.*, 1981) decoupling experiments have proven very useful with respect to the information gained. Figure 6 illustrates the spectral simplification obtainable with these multidecoupling experiments.

E. The Broad-Band Preamplifier

This is a "transparent" accessory for the spectroscopist, except when tuning the unit to the frequency range within which the spectrometer will be pulsing. Broad-band preamplifiers must meet the two following requirements:

(a) large frequency range operation (generally from 2 to 250 MHz);
(b) short recovery time values (generally less than 15 μs) to allow the NMR signal detection of quadrupolar species with corresponding broad lines and very short T_2 values.

F. The Broad-Band Probe Head

The probe head is the key parameter of a multinuclear NMR spectrometer in terms of flexibility and sensitivity.

1. Flexibility

The flexibility can be understood from a listing of the different frequencies that can excite the sample: the lock frequency, the X observing frequency, the proton and (or) Y decoupling frequency, and the X homonuclear decoupling frequency.

In general, because of spatial constraints, two single coils can be wound on the probe insert, and a compromise must be found to keep the probe response free of frequency beats or to degrade too much the Q factor of each resonating circuit. All these requirements lead to a general probe design that uses the decoupler (outer) coil as the lock coil, the inner one being carefully broad-band-adapted to match the desired frequency range via a tuning and a matching circuit. Then homonuclear X decoupling frequency can be injected if necessary into the observing coil through a matched directionnal coupler, which avoids any X homonuclear decoupler frequency breakthrough into the preamplifier.

2. Sensitivity

The requirements for probe sensitivity are very demanding. In fact, in order to obtain good sensitivity, the coupling factor between the observing coil and the sample together with the probe Q factor must be high. On the other hand, in order to recover quickly from the strong (80–150 W) short pulses required to detect fast relaxing nuclei, a low Q is necessary. These conflicting conditions can result in experiments in which high-resolution NMR (liquid state, weak signal) must be done under solid state NMR conditions (very broad lines, short T_2).

3. Acoustic Ringing

If an acoustic ringing response is added with each potential FID, multi-nuclear NMR can become extremely difficult. Because more and more low-γ nuclei are being investigated, it is worth delineating the origin of this phenomenon which can completely obliterate the desired NMR signal.

The source of this spurious signal is the electromagnetic generation of ultrasonic waves in metallic material (Fukushima and Roeder, 1979; Wallace, 1971). This acoustic energy, in the presence of the static magnetic field B_0 is converted into a rf field detected by the coil. Then the rf phase of this acoustic response is coherent with the phase of the exciting pulse, and the amplitude of the spurious signal is given by

$$A = kB_1B_0^2/mv_s(1 + \omega\delta^2/2v_s) \tag{4}$$

where m is the material mass density, v_s the shear velocity of the material, ω the angular frequency, and δ the material skin depth. Hence the acoustic ringing response can be viewed as a broad envelope of frequencies centered around the carrier frequency and in phase with the exciting pulse, namely, the NMR signal. On the other hand, its amplitude is proportional to B_1 (and to the pulse width value), proportional to the square of the static field B_0, and inversely proportional to the pulse frequency (via the ω factor). The acoustic signal occurs generally at about 15 to 20 MHz and below, depending on probe geometry and probe materials. Its duration can reach 2 ms for very low frequencies, which completely obscures any NMR signals broader than 160 Hz. A critical evaluation of the probe components is important to at least attenuate both the amplitude and duration of the signal. A judicious choice of materials, the use of unstrained metallic surfaces, and the replacement of aluminum parts with polymeric materials whenever possible, together with an elaborate coil design and coil attachment on the insert, can reduce both the ringing amplitude by an order of magnitude and its duration to about 100 to 200 μs. Nevertheless, the most promising method for freeing the FID almost completely from acoustic ringing involves a battery of pulse sequences which will be described in Section III.

Section II,A indicated that the multinuclear probe head should deliver values of $\tau^X_{90°}$ as short as possible over the entire frequency range. Actual broad-band emitters and probe heads allow a $\tau^X_{90°}$ of practically constant value (about 40 μs) from ^{109}Ag to ^{31}P.

However, a word of caution is necessary when one has to run several samples with quite different dielectric constants. It is found that the $\tau^X_{90°}$

value determined for a given standard sample at a given temperature increases drastically, especially at high observing frequencies and for solutions with high dielectric constants. A careful retuning of the probe then brings the $\tau^X_{90°}$ value within an acceptable range. The same precaution should be taken when performing variable-temperature experiments.

Another type of probe design used with superconducting magnets is sometimes preferred. It utilizes a solenoid-type coil which results in better intrinsic sensitivity compared to the standard saddle-shaped coil (Hoult and Richards, 1976). Meanwhile, the required orthogonality between the B_1 and B_0 directions requires that the detecting coil be horizontal, which is not convenient for sample handling and magnet shimming. Hence this probe design is mainly used when resolution is not critical.

G. The Variable-Temperature Unit

It is not well recognized that multinuclear NMR observation requires careful temperature equilibration of the sample, because a large family of heavy isotopes may have important $\Delta\delta/K$ factors. Transition metals, for example, have $\Delta\delta\ K^{-1}$ factors between 1 and 3 ppm, which represents an average drift of the resonance of 88 Hz K^{-1} for ^{59}Co at 5.87 T. In fact, this isotope has even been proposed as a sensitive and precise probe thermometer (Levy et al., 1980). In any case, the temperature regulation unit should be switched on continuously to allow $\pm 0.5°C$ accuracy for the probe temperature.

H. The Magnet

A real breakthrough in multinuclear observation, together with the improvement in probe head design, has been the introduction of superconducting magnets.

Apart from the gain in sensitivity they are bringing, proportional to $B_0^{3/2}$, their extreme stability allows overnight accumulation under unlocked conditions—a welcome improvement for the organometallic chemist who often has to prepare a sample under an inert atmosphere and with uncommon solvents. Of course, higher fields imply larger absolute drifts. However, one has to remember that a 10 Hz h^{-1} drift for proton observation at 400 MHz will be lowered to $10(\gamma_X/\gamma_H)/Hz\ h^{-1}$ for isotope X, which represents only 0.4 Hz h^{-1} for ^{183}W, 0.46 Hz h^{-1} for ^{99}Ru, and 0.6 Hz h^{-1} for ^{95}Mo, allowing easy overnight accumulations, provided the sample temperature is regulated.

III. Multinuclear Observation

This section will develop general guidelines governing the multinuclear observation.

A. Sensitivity, Receptivity, and Detectability

At a constant field B_0, the sensitivity of a given isotope X is

$$S_X = K\gamma_X^3 I_X(I_X + 1) \tag{5}$$

The receptivity R_X is expressed as

$$R_X = S_X a_X \tag{6}$$

where a_X represents the X isotope abundance in the solution and a relative receptivity $R_X{}^Y$ can be defined.

In fact, the receptivity factor defined in Eq. (6) must be weighted by a subtle factor $\mathfrak{F}$ which takes into account the following:

(a) the relaxation time T_1 (line width) of the isotope under study in the particular compound dissolved in the NMR tube,

(b) the acoustic ringing amplitude at the operating frequency,

(c) the spectrometer performance in terms of pulse length values and soft- and hardware adaptability. If we consider the signal/noise ratio in terms of spectrometer time compared to the information obtained during a fixed period of time, then, for a given isotope, one can define a detectability factor D_X as

$$D_X = \mathfrak{F}R_X \qquad (\mathfrak{F} \leq 1) \tag{7}$$

This D_X factor governs *all* multinuclear experiments and allows an immediate discrimination between magnetically active isotopes: dipolar ones ($I_X = \frac{1}{2}$) and quadrupolar ones ($I_X > \frac{1}{2}$).

B. Dipolar Isotope Observation

These spin-$\frac{1}{2}$ isotopes can act as very useful probes in mapping out any structural or dynamic features they are involved in. Their intrinsic narrow line widths often allow the detection of J_{XY} couplings. On the other hand, they have long or very long relaxation times T_1 (from 1 to 1 000 s) because of the rarity of abundant dipolar neighbors to speed up the dipole–dipole relaxation pathway. This situation may force the experimentalist to introduce prohibitive interpulse delays T and small pulse angles to satisfy the conditions (Ernst and Anderson, 1966)

$$\cos \alpha_{opt} = e^{-T/T_{1X}} \tag{8}$$

When these recording conditions are coupled with a low receptivity R_X, one can quench the large T_1 values by either dissolving the compound in a protonic solvent or by adding an inert paramagnetic relaxation agent such as $Cr(acac)_3$.

Again, the D_X factor can be increased by recording the X spectrum under continuous ^1H broad-band decoupling to gain from any actual $^1H—X$ nuclear Overhauser effect (NOE) present. However, because of the remote proton environment in organometallic and inorganic molecules, this NOE factor, which theoretically amounts to $\eta_H = \gamma_H/2\gamma_X$ for 100% dipole–dipole relaxation, is generally very low. As many spin-$\frac{1}{2}$ isotopes possess a negative γ (^{109}Ag, ^{107}Ag, ^{29}Si, ^{15}N, ^{103}Rh, ...), a small NOE factor will ineluctably fall into the 0, -1 range, thus decreasing or even nulling the X resonance. Care must be taken to avoid such a situation by using an appropriate recording scheme such as gated decoupling if necessary. Figure 7 provides a good example with the $\{^1H\}—^{29}Si$ spectrum of tetramethylsilane. At room temperature the relaxation of ^{29}Si is mainly governed by spin rotation ($\eta_H = 0$), and ^1H decoupling is effective. As the temperature is lowered, the dipole–dipole relaxation becomes more important and eventually nulls the ^{29}Si signal at 180 K ($\eta_H = -1$). A gated decoupling experiment restores the ^{29}Si signal. As the temperature is lowered further, an inverted $\{^1H\}—^{29}Si$ signal is acquired ($\eta_N^{165K} < -1$).

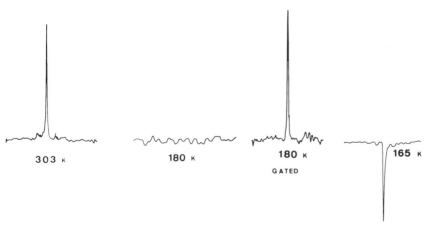

303 K   180 K   180 K
                              G ATED   165 K

Fig. 7   The 17.8-MHz $^{29}Si—\{^1H\}$ spectrum of tetramethylsilane (TMS) recorded at different temperatures. See text for explanation.

A more subtle way to avoid the large T_{1X} value and to obtain a sizable enhancement in sensitivity is to use the INEPT pulse sequence (see Chapter 2), provided that a scalar $^nJ_{X^1H}$ or $^nJ_{X^{31}P}$ coupling exists (Brevard and Schimpf, 1982; Morris and Freeman, 1979). Then the repetition rate is governed only by $T_{1,H(^{31}P)}$, the gain in sensitivity amounting to $\gamma_{H(^{31}P)}/\gamma_X$ for an AX system.

C. Quadrupolar Isotope Observation

Eighty-seven of the 116 magnetically active isotopes are quadrupolar. Apart from the obvious choice between a medium- to high-receptivity spin-$\frac{1}{2}$ isotope and a quadrupolar one for the same element (^{14}N, ^{15}N, ^{199}Hg, ^{201}Hg, ^{129}Xe, ^{127}Xe, etc.), one often has to decide between two quadrupolar nuclides (^{95}Mo, ^{97}Mo, ^{99}Ru, ^{101}Ru, etc.). The line width of such quadrupolar isotopes is given in the extreme narrowing limit by

$$\Delta\nu_{1/2} = \frac{3\pi}{10} \frac{2I_X + 3}{I_X^2(2I_X - 1)} \chi^2 \left(1 + \frac{\eta^2}{3}\right) \tau_c \qquad (9)$$

where χ is the nuclear quadrupole coupling constant defined as $\chi = e^2 q_{zz} Q/h$, Q the nuclear electric quadrupole moment, q_{zz} the largest component of the electric field gradient tensor at the X isotope, e the charge of the electron, η the so-called asymmetry parameter with

$$\eta = (q_{xx} - q_{yy})/q_{zz} \qquad (0 < \eta < 1)$$

and τ_c the isotropic tumbling correlation time of the molecule ($\tau_c \sim 1$–20 ps for nonviscous solutions).

Rearranging Eq. (9) as

$$\Delta\nu_{1/2} = A \frac{2I_X + 3}{I_X^2(2I_X - 1)} Q^2 \tau_c \qquad (10)$$

clearly indicates that the choice between two quadrupolar isotopes of the same element to start NMR observation will be governed by the "isotope IQ" factor $(2I + 3)Q^2/I^2(2I - 1)$; the lower this IQ factor, the sharper the resonance line, hence the better the detectability D_X.

On the other hand, once the best quadrupolar isotope has been chosen, the χ^2 term will be governing the resonance width from compound to compound; this line width can vary from 1 Hz to more than 1 kHz, depending on the electric field gradient at the observed nuclide.

When such a broad line is expected, the only way to sharpen it is to decrease the molecular tumbling time τ_c by using a nonviscous solvent and recording the spectrum at the highest temperature allowed by both the solvent boiling point and the solute stability.

Of course, no INEPT-like sequences are available for quadrupolar isotopes, as the quadrupolar relaxation (see Chapter 5) acts as an efficient decoupler for any $J_{XH(^{31}P)}$ scalar coupling except for very symmetric electronic environments which are seldom encountered.

D. Elimination of Acoustic Ringing

As explained in Section II,F,3, all low-frequency NMR experiments have this problem. In fact, one can consider subtracting a blank spectrum from the normal one by accumulating the same number of scans on a tube filled with solvent only or by shifting the carrier offset several kilohertz away. Apart from doubling the experimental time, these methods seem to result in no suppression of acoustic ringing at all, because they introduce a modification of the ringing response of the probe between the blank and the desired spectrum.

Fortunately, the characteristics of this spurious acoustic signal have initiated the devising of some new pulse sequences that lower the ringing detection within quite reasonable limits.

A straightforward solution is to insert a delay of DE microseconds between the end of the pulse and the start of the acquisition. The DE value should be less than $1/\pi\Delta\nu_{1/2,X}$, otherwise this delay will also cut off a noticeable part of the FID. The method then holds for sharp lines (dipolar or symmetric quadrupolar species), but it adds an extra dephasing process which may render a multiline spectrum quite difficult to phase via the first- and second-order phase correction routines of the computer.

A more appropriate solution is the use of pulse sequences. Indeed, as the acoustic ringing response is in phase with the exciting pulse, and because its intensity depends on the length of this pulse, one can imagine a pulsing scheme that subtracts sequentially a pair of FIDs, where

$$\text{FID}_1 = \text{NMR signal} + \text{ringing}$$
$$\text{FID}_2 = -\text{NMR signal} + \text{ringing}$$

Inversion of the NMR signal in FID_2 is obtained via a 270° pulse or a judicious choice of rf phase shifting within the sequence itself.

The following sequences have been used to eliminate the strongest acoustic response; Δ is a delay ranging from 1 to 100 μs.

$$[(90°_x\text{-AQT})_3\text{-}(270°_{-x}\text{-AQT})]_n \tag{11}$$

$$[(90°_x\text{-}\Delta\text{-}180°_y\text{-}\Delta - \text{AQT})\text{-}(90°_{-x}\text{-}\Delta\text{-}180°_y\text{-}\Delta\text{-}\text{AQT})]_n$$

(Brevard and Schimpf, 1981)   (12)

$$[(90°_x\text{-}\text{AQT})\text{-}(180°_x\text{-}\Delta\text{-}90°_x\text{-}\text{AQT})]_n$$

(Canet *et al.*, 1982)   (13)

Finally, a most efficient pulse sequence has been recently introduced (Ellis, 1982). It is written

$$D_1\text{-}180°_{\phi_1}\text{-}\alpha°_{\phi_2}\text{-}D_2\text{-}180°_{\phi_3}\text{-}D_3 - \text{acquire FID with receiver phase } \phi_4$$
$$D_1\text{-}\alpha°_{\phi_1}\text{-}D_2\text{-}180°_{\phi_3}\text{-}D_3 - \text{acquire FID with receiver phase } \phi_4$$
(14)

and repeating the sequence n times, where α is a particular flip angle (90° or less) with associated phase $\phi1,2$, 180° the nominal flip angle with associated phase $\phi1,3$, D_2 and D_3 the delay times between 1 and 100 μs according to the initial acoustic ringing amplitude [generally $D_3 < D_2$ for

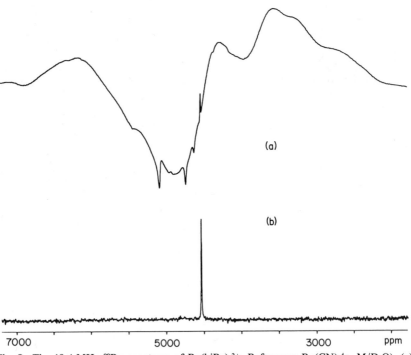

Fig. 8 The 18.4-MHz ^{99}Ru spectrum of Ru(biPy)$_3^{2+}$. Reference: Ru(CN)$_6^{4-}$, M/D$_2$O). (a) Normal acquisition with 5-μs delay between end of pulse and acquisition. (b) Acquisition with sequence (14), $D_2 = D_3 = 5$ μs, $D_1 = 0$ s, and $\alpha = 90°$.

sampling the FID at the echo maximum, but $D_3 = D_2$ works quite well (Fig. 8)], D_1 a relaxation delay according to T_{1X}, ϕ_4 the receiver phase during each FID sampling, and ϕ_1, ϕ_2, and ϕ_4 are written as

$$\phi_1: \quad xxxx \ \bar{x}\bar{x}\bar{x}\bar{x} \ yyyy \ \bar{y}\bar{y}\bar{y}\bar{y}$$

$$\phi_2: \quad \bar{x}\bar{x}\bar{x}\bar{x} \ xxxx \ \bar{y}\bar{y}\bar{y}\bar{y} \ yyyy$$

$$\phi_3: \quad y\bar{y}x\bar{x} \ y\bar{y}x\bar{x} \ \bar{x}x\bar{y}y \ \bar{x}x\bar{y}y$$

$$\phi_4: \quad RyRy\overline{RyRy} \ \overline{RyRyRyRy} \ \overline{RxRxRxRx} \ RxRx\overline{RxRx}$$

The total number of accumulated FIDs consists of $2n$ blocks of 16 FIDs each to allow for complete phase cycling over the entire phase program.

Figure 8 illustrates the efficiency of the sequence in eliminating almost perfectly any ringing response.

References

Bax, A., Freeman, R., and Kempsell, S. P. (1980a). *J. Am. Chem. Soc.* **102**, 4849.

Bax, A., Freeman, R, and Kempsell, S. P. (1980b). *J. Magn. Reson.* **41**, 349.

Bendall, M. R., Doddrell, D. M., and Pegg, D. T. (1981). *J. Magn. Reson.* **44**, 238.

Bodenhausen, G., Freeman, R., and Morris, G. A. (1976). *J. Magn. Reson.* **23**, 171.

Brevard, C., and Schimpf, R. (1981). Unpublished results.

Brevard, C., and Schimpf, R. (1982). *J. Magn. Reson.* **47**, 528.

Canet, D., Brondeau, J., Marchal, J. P., and Robin-Lherbier, B. (1983). *Org. Magn. Reson.* In press.

Ellis, P. D. (1982). NATO Summerschool on "multinuclear approach to NMR spectroscopy", August 1982, Stirling, Scotland.

Ernst, R. R., and Anderson, W. A. (1966). *Rev. Sci. Instrum.* **37**, 93.

Fukushima, E., and Roeder, S. B. W. (1979). *J. Magn. Reson.* **33**, 109.

Granger, P., and Schrobilgen, G. J. (1982). NATO Summerschool on "multinuclear approach to NMR spectroscopy", August 1982, Stirling. Scotland.

Harris, R. K., Knight, C. T. G., and Hull, W. E. (1981). *J. Am. Chem. Soc.* **103**, 1577.

Heaton, B. T., Strona, L., Martinengo, S., and Chini, P. (1980). *J. Organomet. Chem.* **194**, C29.

Hoult, D. I., and Richards, R. E. (1976). *J. Magn. Reson.* **24**, 71.

Levy, G. C., Bailey, J. T., and Wright, D. A. (1980). *J. Magn. Reson.* **47**, 353.

Morris, G. A., and Freeman, R. (1978). *J. Magn. Reson.* **29**, 433.

Morris, G. A., and Freeman, R. (1979). *J. Magn. Reson.* **101**, 761.

Van Stein, G. C., Van Koten, G., and Brevard, C. (1982). *J. Organomet. Chem.* **226**, C27.

Wallace, W. D. (1971). *Int. J. Nondestructive Test* **2**, 309.

2

Techniques That Enhance Sensitivity, Improve Resolution, Correlate NMR Spectral Parameters, and Lead to Structural Information

Philip H. Bolton

Department of Chemistry
Wesleyan University
Middletown, Connecticut

I. Introduction

During the past decade or so the methodology of NMR has undergone considerable change. Following and preceding chapters deal with the advantages of high-field spectrometers, better probe designs, and the coupling of computers with pulsed Fourier transform (FT) experiments. These technological improvements have greatly increased the sensitivity and resolution of NMR. In this chapter experimental techniques that enhance sensitivity or resolution and aid in the making of assignments will be presented and explained.

The increase in sensitivity observed with increasing magnetic field

Copyright © 1983 by Academic Press, Inc.
All rights of reproduction in any form reserved.
ISBN 0-12-437101-9

strength has been known for some time (Ernst, 1966). However, a more dramatic increase in the sensitivity of the detection of low-gyromagnetic-ratio nuclei can often be obtained by the transfer of magnetization from a higher gyromagnetic ratio nucleus. The resolving power of spectrometers often increases with increasing magnetic field strength. A much greater gain in resolution can be obtained by a spreading out of the spectral information in two dimensions. The ability to present the chemical shifts of ^{11}B along one axis and the heteronuclear couplings along an orthogonal axis, for example, brings an increase in resolving power to NMR much greater than that obtained by the ever-increasing field strengths achieved by manufacturers of NMR spectrometers. Correlation of the chemical shifts of different nuclei belonging to the same coupling network has superseded the use of many decoupling and triple-resonance experiments. Simultaneous correlation of the chemical shifts of many pairs or triples of nuclei offers a considerable saving in spectrometer time and hence can be thought of as increasing the sensitivity.

II. Sensitivity Enhancement Techniques

The sensitivity of any experiment depends on the strength of the signal actually observed. In FT NMR the signal is the current induced in a coil by a precessing magnetization vector. At equilibrium in a large static magnetic field a nucleus I, with spin $\frac{1}{2}$, has two energy levels separated by

$$\Delta E = \gamma_I B_0$$

where γ_I is the gyromagnetic ratio of nucleus I and B_0 the strength of the applied field. Since ΔE is very small compared with kT for any liquid, it can be assumed that Boltzmann statistics are applicable. Thus the populations of the two energy levels are

$$P_\beta/P_\alpha = e^{-\Delta E/kt} = e^{-\gamma_I B_0/kT}$$
$$P_\beta = \frac{1}{2}(1 - \gamma_I B_0/kT)N \qquad P_\alpha = \frac{1}{2}(1 + \gamma_I B_0/kT)N$$

where N is the total number of I nuclei.

The net magnetization along the z axis is the population difference between "up" and "down" orientations of the nuclei times the magnetic moment of the nucleus. Thus the net magnetization along this axis at equilibrium depends on γ_I^2.

In a FT experiment a short rf pulse is used to rotate the magnetization from the z axis into the xy plane. The magnetization vector then freely precesses in the xy plane at the Larmor frequency of the nucleus:

$$\text{Larmor frequency} = \gamma_I B_0$$

The strength of the induced current depends on the magnitude of the magnetization vector times the Larmor frequency (remember that $\Delta E = h\nu$ tells us that increasing the frequency increases the strength of the induced signal). Therefore the sensitivity of the NMR experiment is dependent on γ_I^3 (Ernst, 1966). The receptivity of a nucleus is typically given as

$$R_I \equiv (\gamma_I/\gamma_C)^3[S_I(S_I + 1)/S_C(S_C + 1)]$$

where S_I is the spin of nucleus I ($\frac{1}{2}$, 1,...) and γ the gyromagnetic ratio of ^{13}C which has a standard receptivity R of 1 and a spin of $\frac{1}{2}$.

The spin of the nucleus affects the net magnetization but not the precessional frequency. The receptivity of ^{109}Ag, for example, whose gyromagnetic ratio is one-fourth that of ^{13}C, is 0.006 for equal numbers of nuclei.

Low-gyromagnetic-ratio nuclei have a very low sensitivity, requiring signal averaging to improve the signal/noise ratio. In the limit that the noise is perfectly random, or white, the signal/noise ratio improves with the square root of the number of the independent data sets. Therefore ^{109}Ag requires $\sim 4 \times 10^5$ times as long as a ^{13}C spectrum. If a ^{13}C spectrum takes 1 s to acquire, than a ^{109}Ag spectrum will require approximately 1 month.

An additional problem with low-gyromagnetic-ratio nuclei with spin $\frac{1}{2}$ is that the relaxation times tend to be very long, making the situation even more distressing. These factors, coupled with the low natural abundance of ^{15}N and other nuclei of potential interest, tend to portray the utility of the NMR of such nuclei as almost nil.

The basic instrumentation features for increasing sensitivity are discussed in Chapter 1. However, there are always limits on instrumentation. What is needed are basic changes in *experimental* procedures for enhancing the detected signal.

The preceding discussion points out ways in which the signals of low-gyromagnetic-ratio nuclei might be enhanced. One method is to increase population differences before the observation pulse by coupling the low-gyromagnetic-ratio nucleus population difference to that of a high-sensitivity nucleus. The nuclear Overhauser effect (NOE), magnetization transfer, and cross-polarization techniques all use this approach. A real benefit of the magnetization transfer technique is that the recycle time of the experiment is limited by the relaxation rate of the high-sensitivity nucleus. Another method is to detect the low-gyromagnetic-ratio nuclei via a more sensitive nucleus, which typically involves two-dimensional spectroscopy. As shown in the following discussion, some of these experiments can offer truly spectacular gains in sensitivity in actual use.

III. The Nuclear Overhauser Effect

The NOE exploits the dipolar coupling to increase the populations between energy levels connected by transitions of a low-gyromagnetic-ratio nucleus and is typically applicable only to spin-$\frac{1}{2}$ nuclei. For quadrupolar nuclei the relaxation is typically not predominately dipolar, and there is no NOE.

The origin of the NOE can be illustrated for the case of two nuclei that are dipolar- but not scalar-coupled to one another. Figure 1 shows the energy levels for such a spin system and indicates the transitions of interest. As usual, the two nuclei are designated I and S. This designation is based on Ionel Solomon's initials (1955). In this example the S nuclei have the larger population difference at equilibrium.

The basic idea behind the NOE is to couple the larger ΔE of the S spins to the smaller ΔE of the I spins to increase the population differences between the energy levels connected by I transitions. The coupling between the populations of the two nuclei is the dipolar relaxation indicated in Fig. 1.

The intensity of a NMR transition is given by

$$I = P_\beta - P_\alpha, \qquad I_1 = (P_3 - P_1) + (P_4 - P_2)$$

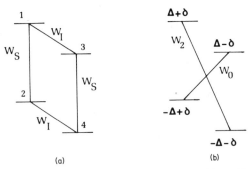

(a)   (b)

Fig. 1 (a) The energy levels of an IS spin system in which I and S are dipolar-coupled to one another. The I spin transitions connect energy levels 1 and 3 as well as 2 and 4, whereas the S transitions connect levels 1 and 2 in addition to 3 and 4. The rates of relaxation due to single spin flips are indicated by W_I and W_S which indicate relaxation due to a flip of the I or S spin, respectively. (b) The concerted two-spin transitions. For the rate W_0, the I and S spins going between states are represented by $\alpha\beta$ and $\beta\alpha$, with α and β indicating the polarizations of the two spins. The rate W_2 corresponds to the transition $\alpha\alpha \leftrightarrow \beta\beta$.

At equilibrium the populations of the four energy levels are (disregarding a common multiplicative factor)

$$P_1 = 1 - p - q, \qquad P_3 = 1 - p + q$$
$$P_2 = 1 + p - q, \qquad P_4 = 1 + p = q$$

where $p = \gamma_I B_0 / 2kT$ and $q = \gamma_S B_0 / 2kT$. Thus, at equilibrium

$$I_1 = (1 - p + q) - (1 - p - q) + (1 + p + q) - (1 + p - q) = 4q$$

When the S transitions are saturated and there is no relaxation, the populations of the energy levels are

$$P_2 = P_1, \qquad P_3 = P_4 \qquad \text{(due to saturation)}$$
$$P_2 = [(1 - p - q) + (1 + p - q)]/2,$$
$$P_3 = [(1 - p + q) + (1 + p + q)]/2,$$

which simplifies to

$$P_1 = P_2 = (2 - 2q)/2, \qquad P_3 = P_4 = (2 + 2q)/2$$

In the limit of no relaxation the intensity of the I signal is

$$I_1 = 4q$$

which is unchanged from the equilibrium value, implying that relaxation is required for a change in the intensity of the I signal. When the S spin is saturated and W_2 is much greater than any other rate of relaxation, P_1 and P_4 tend toward their equilibrium population difference:

$$P_1 \longrightarrow 1 - p - q, \qquad P_2 = 1 - q$$
$$P_4 \longrightarrow 1 + p + q, \qquad P_3 = 1 + q$$

The intensity of the I transition differs from the equilibrium value:

$$I_1 = (1 + q) - (1 - p - q) + (1 + p + q) - (1 - q) = 2p + 4q$$

The increase in the I signal intensity is half the ratio of the gyromagnetic moments of the I and S nuclei.

$$I_1 = [\tfrac{1}{2}(\gamma_S/\gamma_I) + 1]4q$$

When $W_2 \simeq W_0$ and W_1 is negligible, P_2 and P_3 tend toward their equilibrium difference due to W_0:

$$P_2 \longrightarrow 1 + p - q, \qquad P_3 \longrightarrow 1 - P + q$$

Because W_2 relaxation is also occurring, P_1 and P_4 also tend toward their equilibrium difference:

$$P_1 \longrightarrow 1 - p - q, \qquad P_4 \longrightarrow 1 + p + q$$

The intensity of the I-nucleus signal is again that obtained at equilibrium.

$$I_I = (P_3 - P_1) + (P_4 - P_2)$$
$$= (1 - p - q) - (1 - p + q) + (1 + p + q) - (1 + p - q)$$
$$= 4q$$

Therefore, the NOE is useful only for enhancing I signal intensities when $W_2 \gg W_0$. In practice, this means that the NOE works only for relatively small molecules. It has been shown (Noggle and Shirmer, 1971) that, for a molecule with a single correlation time,

$$f = \frac{W_2 - W_0}{2W_I + W_2 + W_0} \frac{\gamma_S}{\gamma_I} = \text{fractional enhancement due to NOE}$$

where

$$W_2 = \frac{3}{5} \frac{\gamma_I^2 \gamma_S^2}{r^6} \frac{\tau_c}{1 + (W_I + W_S)^2 \tau_c^2}$$

$$W_0 = \frac{1}{10} \frac{\gamma_I^2 \gamma_S^2}{r^6} \frac{\tau_c}{1 + (W_I - W_S)^2 \tau_c^2}$$

$$W_I = \frac{3}{20} \frac{\gamma_I 2 \gamma_S^2}{r^6} \frac{\tau_c}{1 + W_I^2 \tau_c^2}$$

and r is the distance between the I and S nuclei.

When the molecular motion cannot be described by a single correlation time, as for many proteins, polymers, and polynucleotides, more complicated expressions for the NOE are obtained (Allerhand *et al.*, 1972; Bolton and James, 1980). The key feature is that the NOE is primarily sensitive to the fastest molecular motion. That is, even if the molecule of interest is undergoing slow tumbling in solution, an appreciable NOE can exist if there is rapid local or internal motion.

Disadvantages of the NOE are that it does not work for molecules of moderate size, because W_0 becomes comparable to or greater than W_2. The recycle time of the experiment is related to the relaxation of the I nuclei, which can be long for small molecules. In addition, in even the most favorable instances only half of the population difference of the sensitive nucleus can be transferred to the I spins. Another disadvantage is that, for nuclei with a negative gyromagnetic ratio, the NOE can be -1, hence *no signal* is observed. This is a rather common occurrence for ^{15}N, for example. Because of these problems direct magnetization transfer from S to I spins was developed and is presented next.

IV. Sensitivity Enhancement via Population Transfer

Enhancement of the signals of insensitive nuclei can utilize the scalar, spin–spin rather than the dipolar coupling as the linkage for transfer of magnetization from sensitive to insensitive nuclei. This approach is not limited to spin-$\frac{1}{2}$ nuclei but, in general, requires that the heteronuclear couplings be resolvable.

The selective population transfer (SPT) experiment can be illustrated by considering a simple IS spin system. The energy levels are shown in Fig. 2 along with the spectra of the I and S nuclei. The populations of the energy levels are perturbed by inducing transitions of the S nuclei and observed via the I nuclei.

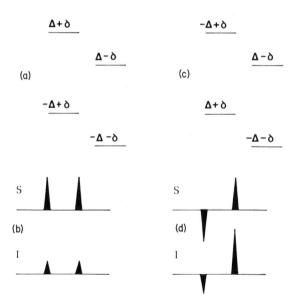

Fig. 2 (a) The energy levels and populations of a spin–spin, scalar-coupled IS spin system. The numbering of the energy levels and the types of energy level connections are as in Fig. 1. There is a population difference of 2Δ between energy levels connected by S transitions and 2δ between energy levels connected by I transitions at equilibrium. (b) The intensities and frequencies of the I and S signals obtained for the populations in (A). (c) Interchange of the populations of energy levels 1 and 2 leads to the populations shown. The inversion can be accomplished either by a selective 180° pulse applied at the frequency of the 1,2 transition or by the INEPT pulse sequence. (d) The nonequilibrium population distribution in (c) leads to the unusual intensities but unchanged frequencies of the I and S signals, as shown.

The SPT experiment uses a selective pulse to invert the populations of the energy levels connected by one of the two S transitions. In the example shown in Fig. 2 the populations of energy levels 1 and 2 are interchanged by a selective 180° pulse applied at the frequency of the 1,2 transition, which is analogous to the inversion of signals in a T_1 experiment. After the population redistribution the populations are monitored via a normal observe pulse applied to the I nuclei. As indicated in Fig. 2, the I signals have the usual frequencies but unusual intensities, the gain in intensity being the ratio of the gyromagnetic ratios of the I and S nuclei. Thus the SPT experiment can have twice the enhancement of the NOE. A considerable advantage of the SPT experiment is that it can be repeated when the sensitive nucleus magnetization recovers, since the population differences observed as I magnetization originate as sensitive nucleus magnetization. Because for many nuclei of interest the sensitive nucleus T_1 can be much less that the I spin T_1, the rapid recycle time is an attractive feature.

The SPT experiment as just described has two drawbacks which can be overcome. The first is that only one heteronuclear pair can have its populations properly redistributed in one experiment because of the nature of the selective inversion. The other problem is that there is no *net* enhancement. This means that, if the I signal is acquired with S spin decoupling, then there is no gain in signal intensity.

The selectivity of the SPT experiment can be overcome by the use of a pulse sequence, known as INEPT, which inverts the populations across the energy levels connected by only one of the two S spin transitions regardless of the S spin chemical shift (Freeman and Morris, 1979b; Morris, 1980). This procedure is illustrated in Fig. 3. Once the populations have been properly redistributed, the I pulse is applied. The signals obtained in SPT and INEPT experiments exhibit no net enhancement. This second problem can be simply overcome as follows. The two I magnetizations are of opposite algebraic sign. The INEPT procedure takes the S magnetizations that were originally in phase with respect to J_{IS} and brings them into phase opposition. Thus the I magnetizations are brought into phase with one another by running the INEPT sequence again (Burum and Ernst, 1980; Bolton 1980). This involves refocusing of the I chemical shifts so that the final spectra have all signals of the same algebraic sign and has been dubbed INEPTR, INEPT with refocusing.

The giving of acronyms to pulse sequences, a popular practice nowadays, was apparently started by John Waugh when he named a sequence WAHUHA, allegedly after the authors of a technique for suppressing homonuclear dipolar coupling, who were Waugh, Huber, and Haeberlen (Waugh *et al.*, 1968). In actual fact WAHUHA is also the

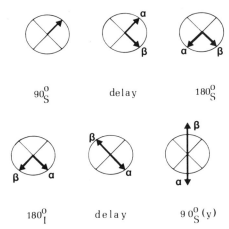

$$90^0_S \qquad\qquad delay \qquad\qquad 180^0_S$$

$$180^0_I \qquad\qquad delay \qquad\qquad 90^0_S(y)$$

Fig. 3   The behavior of the S magnetization vectors during an INEPT pulse sequence. The 90° pulse applied to the S spins rotates the z magnetization into the xy plane. During the delay time equal to $1/4 J_{IS}$ the two S magnetization vectors precess away from one another. The difference in the precessional frequencies is simply J_{IS}. After the delay time simultaneous, or almost simultaneous, 180° pulses are applied to the I and S spins. The 180° S pulse rotates the S spin magnetization vectors 180° about the y axis as shown. The 180° I spin pulse interchanges the labels, α and β, of the S magnetization vectors. This interchanging of labels is due to interconversion of the I spin α and β. The S magnetization vectors are then allowed to precess for a time $1/4 J_{IS}$ once more, and at the end of the delay are in antiphase with one another and lie along the x axis. A 90° pulse applied to the S spins whose alignment is along the y axis results in inversion, pointing straight down, of only one of the S magnetization vectors. Thus selective inversion of one of the S magnetizations has been accomplished with no dependence on the chemical shift of the S spins. For more details see Bolton (1980), Burum and Ernst (1980), Freeman and Morris (1979b), and Morris (1980).

school cheer of Dartmouth University where Waugh was an undergraduate. The naming of pulse sequences is now a routine procedure of which this author is also guilty.

INEPT and INEPTR have been applied to a number of samples and are a common feature in the brochures of instrument manufacturers. An illustration of the advantage of the INEPTR method is detection of the ^{15}N signal of formamide as shown in Fig. 4. When there is a short recycle time for the experiment, the INEPTR method gives a signal intensity about seven times greater than that obtained using the NOE, which reduces the time for equivalent signal/noise spectra 50-fold. When the recycle time is much longer than either the ^{15}N T_1 or the ^{1}H T_1, then the intensity obtained by the INEPTR method is only about 150% that of the NOE approach.

An analogous procedure for enhancement of the signals of insensitive nuclei utilizing the scalar coupling as the linkage is cross-polarization

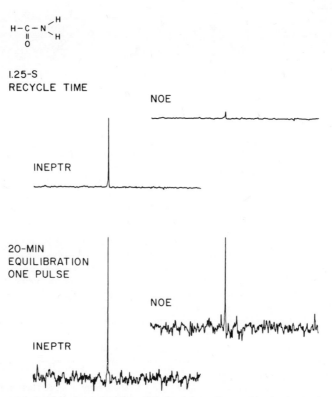

Fig. 4 Comparison of the intensity of the ^{15}N signal from formamide obtained using NOE or INEPTR as the method of transferring magnetization from ^{1}H to ^{15}N. The data are from Bolton (1980).

(Chingas *et al.,* 1979, 1980). In this type of experiment the proton and insensitive nuclei magnetizations are first rotated into the *xy* plane. After the initial rotation both spins are subjected to "spin-locking" fields, such that the protons and insensitive nuclei are precessing in a rotating frame at the same frequency. This is accomplished by applying spin-locking fields to the two nuclei, the ratio of the field strengths applied to the two nuclei being the inverse of their gyromagnetic ratios. Magnetization is passed from S to I nuclei in an oscillatory manner with a frequency equal to the heteronuclear coupling. The cross-polarization method can give results equal to that of INEPT-type experiments but is typically feasible only with a specially constructed spectrometer that has very good homogeneity of the spin-locking fields. With commercially available spectrometers the INEPTR experiment is much easier to implement successfully.

There are, however, two possible problems with INEPTR and cross-polarization experiments. Selective inversion of the S populations can be carried out quite successfully when the S spins are coupled only to a single I nucleus, hence the same timing is essentially optimal for all S spins. However, the I nucleus can be coupled to three or more S spins. This implies that a compromise choice for bringing all the I magnetization vectors into phase must be made. The case of ^{13}C (with one, two, or three attached protons) has been studied, and not much is lost by making a compromise choice (Burum and Ernst, 1980; Bolton 1980). Because the real advantage of the INEPTR experiment is in the short recycle time, a slight deviation from optimal magnetization transfer does not seriously detract from the utility of the approach.

The other limitation of INEPTR, and of all other population transfer methods, is molecular size. For large molecules, greater than ~20,000 daltons, the T_1 of the I nuclei, such as ^{13}C, can become shorter than the S, or ^1H, T_1. This obviates the recycle time advantage found for small molecules. In addition, there can be overlap between the S lines split by the heteronuclear coupling, and this reduces the specificity of the original redistribution of the populations. In actual practice it has been found that the INEPTR approach is not superior to the NOE technique when applied to ^{13}C detection of the signals from tRNA which has a molecular weight of about 25,000 (Bolton, 1980).

In some cases of interest the I-nucleus signal need not be detected directly. For example, if one is interested only in heteronuclear coupling, an S spin observation INEPT experiment can be used. This procedure allows the detection of only the S nuclei coupled to the heteronucleus of interest (Freeman et al., 1981), hence heteronuclear couplings are detected with high sensitivity.

A more elaborate version of this technique has been applied in the detection of ^{15}N signals in natural abundance in 5-mm tubes (Bodenhausen and Ruben, 1980). The first step is a standard INEPT magnetization transfer from ^1H to ^{15}N. The magnetization is then allowed to precess at the ^{15}N chemical shift and finally transferred back to the protons for observation (Bodenhausen and Ruben, 1980). This two-dimensional experiment (see the following discussion for more details) in principle allows enhancement of the signal from ^{15}N by $(\gamma_H/\gamma_N)^2$. The power 2 of the gyromagnetic ratio comes from the initial population transfer together with observation at the ^1H (Larmor frequency) rather than the ^{15}N.

A slight variation on this theme has been used for indirect detection of ^{199}Hg signals (Bodenhausen and Roberts, 1980; Vidusek et al., 1982). The chemical shift range for ^{199}Hg is very large, and identification of the ligands can often be based on an approximate determination of the ^{199}Hg chemical shift. The amino acid-binding sites of methyl mercury have been investigated by determining the chemical shift of the ^{199}Hg in about 10-kHz increments via proton observe INEPT with frequency selectivity provided by the spectral window of the pulses applied to the ^{199}Hg spins.

The method of choice for enhancement of the signals of low-gyromagnetic-ratio nuclei is the INEPT experiment. This experiment is easily implemented on any modern NMR spectrometer and is relatively insensitive to the precise setting of the pulse widths and times. The method is limited to nuclei that are scalar-coupled to protons or other high-gyromagnetic-ratio nuclei [an example of enhancing the sensitivity of ^{103}Rh via magnetization transfer from ^{31}P has been presented (Brevard and Schimpf, 1982)] and to molecules of small to moderate size. However, the advantages of transferring the entire S spin population difference to the I nucleus and exploiting the relatively rapid relaxation of the protons make the INEPT experiment clearly superior to the NOE method in a wide range of applications.

V. How to Determine the Number of Protons Directly Coupled to a Heteronucleus

A determination of the number of protons directly coupled to a heteronucleus can be obtained from a proton-coupled spectrum. In many cases of interest the straightforward method fails because of extensive overlap of the signals from different I nuclei. In the case of ^{13}C, which is typically thought to have a wide distribution of chemical shifts, the proton-coupled spectrum of even relatively small molecules can be complicated and difficult to interpret.

A number of methods have been proposed for scaling the heteronuclear couplings to overcome the overlapping of signals. One approach that was popular for a number of years was the use of continuous-wave, coherent proton irradiation which is not on-resonance with the proton signals (Anderson and Freeman, 1962). Off-resonance decoupling reduces the heteronuclear couplings, hence reduces overlap, but has serious drawbacks and is not recommended (Freeman et al., 1978).

Another approach uses selective excitation of a single multiplet (Morris and Freeman, 1978). This is accomplished by performing the selective excitation while the sample is decoupled and then acquiring the coupled signal of the selectively excited site. This method overcomes the spectral distortions of the off-resonance experiment but is necessarily very time-consuming because each multiplet requires the same time as a coupled spectrum. That is, a sample with 10 sites whose multiplicities are unknown requires a time equal to that needed to acquire 10 coupled spectra. The desired experiment would allow determination of the multiplicities of all the sites in a single experiment and not take much more time than a conventional decoupled spectrum.

The most reliable approach to determination of the number of protons directly bonded to a heteronucleus is heteronuclear two-dimensional J spectroscopy (Freeman *et al.*, 1977; Müller 1979; Müller *et al.*, 1977). This provides spectra that have the chemical shift of the heteronucleus along one axis and the proton-coupled spectrum along a perpendicular axis.

The basic idea of J spectroscopy is to allow the I nuclei to precess while coupled to the protons during an evolution time and then to acquire the proton-decoupled spectrum. The intensities of the proton-decoupled signals depend on the length of the evolution time, the multiplicity, and the size of the coupling.

A vector model is sufficient for description of the behavior of the I spin magnetization vectors in a two-dimensional J experiment in the absence of strong coupling among the protons (Freeman *et al.*, 1977). When all the heteronuclear couplings are the same, then the modulation of the intensities of different multiplicities is given by

$$IH(t_1) = A \cos \theta$$
$$IH_2(t_1) = A/2(1 + \cos 2\theta)$$
$$IH_3(t_1) = A/4(3 \cos \theta - \cos 3\theta)$$

where $\theta = 2\pi J_{IH} t_1$ and A is the unmodulated intensity.

Thus Fourier transformation of the modulation of the intensities of the I signals with respect to t_1 gives two signals for an IH at $\pm J/2$, three signals for an IH$_2$ at $\pm J$ and 0, and four signals for an IH$_3$ at $\pm 3J/2$, $\pm J/2$. A typical example is shown in Fig. 5. It is a simple matter to determine the multiplicities of the different sites from the J spectrum. The time required for a J spectrum is typically on the same order as for a T_1 or a chemical shift correlation experiment. Even though J spectra can have a very high resolution, because the line widths are not limited by the field homogeneity, in a typical application where only the multiplicities are of interest, quite a low resolution in the J dimension is sufficient, as shown in the contour map of Fig. 5.

The two-dimensional J approach works regardless of the magnitude of J or the multiplicity. Its sole drawback is the time required for the experiment. This drawback has led to the development of one-dimensional analogs of J spectroscopy to determine the multiplicity from the data in one or two spectra. The basis of the one-dimensional experiments is that the

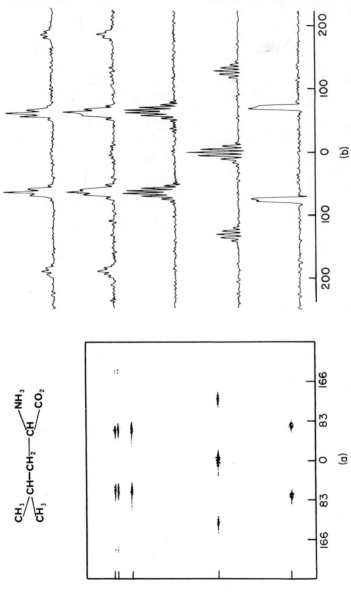

Fig. 5 Heteronuclear two-dimensional J spectra of leucine shown as a contour map. Leucine has five distinguishable carbon sites with two distinct methyl signals because there is a prochiral center. In the contour map the ^{13}C chemical shifts are along the vertical axis and the J coupling along the horizontal axis. The pulse sequence used to obtain the data shown was the proton flip method (Freeman and Morris, 1979a). This pulse sequence consists of a $90°_I - t_1/2 - 180°_S - 180°_I - t_1/2$-acquire I signal with S decoupling. The bottom signal is from the α carbon site and clearly shows that it is coupled to only one proton because the J spectrum is a doublet. Similarly, one can readily determine the multiplicity of all five carbon sites from the contour map. (b) The same data as in (a) are shown in the phase-sensitive mode. The resolution is much higher than in the contour map mode, and long-range ^1H—^{13}C couplings are discernible.

intensities of the IH, IH_2, and IH_3 signals are different for different values of t_1 as just discussed. Thus, by performing the experiment for judiciously chosen values of t_1, the different kinds of sites can be distinguished and perform quite well. However, the heteronuclear coupling is not the same for all sites. For example, a typical $^1J_{CH}$ coupling for sp^3 hybridization is 125 Hz, for sp^2 hybridization 155 Hz, and for sp hybridization 240 Hz. The heteronuclear coupling also depends on the ligands, $^1J_{CH}$ being 149 Hz for CH_3F and 239 Hz for CHF_3.

In the converse case, when the molecule of interest has only one kind of hybridization and no unusual ligands, the experiment is straightforward, with typical results shown in Fig. 6. The experiment can distinguish between CH and CH_3 as one group and CH_2 as a separate group in one determination (Burum and Ernst 1980). To distinguish between CH and CH_3 the delay time used is chosen to maximize the difference between these two. For a particular J the best choice is to have $\theta \simeq 54.7°$ (Jakobsen et al., 1982). Figure 6 shows data obtained using such a procedure. Thus,

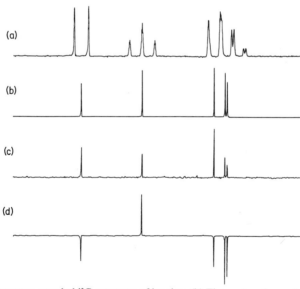

Fig. 6 (a) The proton-coupled ^{13}C spectrum of leucine. (b) The proton-decoupled spectrum of leucine. Even for this relatively simple example there is extensive overlap of the signals in the proton-coupled spectrum, making interpretation difficult. The spectrum in (d) was obtained using a $90°_I$–delay–$180°_S$–$180°I$–delay-acquire I signal with an S spin-decoupled pulse sequence. The delay was set equal to $1/J$ which allows maximal differentiation between CH_2 and the CH and CH_3 signals. To distinguish between the CH and CH_3 signals the delay time is set to the "magic angle" as described in the text and the spectrum in (c) is obtained. It is noted that the signals from the CH_3 sites are significantly reduced relative to the signals from CH sites, allowing their identification.

in favorable cases when all the $^1J_{\text{IH}}$ are very similar, just one or two one-dimensional experiments can allow determination of the multiplicities of all the sites on a molecule. However, one must always be cautious and keep in mind that the couplings can vary over a wide range and that one-dimensional experiments are limited by the assumption of a single $^1J_{\text{CH}}$.

In addition to the method presented here at least two other approaches have been presented, which are known as distortionless enhancement by polarization transfer (DEPT) (Doddrell *et al.*, 1982) and the attached proton test (APT) (Patt and Shoolery, 1982). These methods give results essentially equivalent to those of the approach previously described, and all three techniques have limitations because they share the same underlying basic feature. There is not much on which to base a choice among the three methods except ease of implementation on a particular spectrometer. All three types of experiments can be performed with any modern spectrometer using nuclei other than ^{13}C. The sensitivity for any single choice of t_1 is comparable to that of a conventional decoupled spectrum (Doddrell *et al.*, 1982; Jakobsen *et al.*, 1982; Patt and Shoolery, 1982).

VI. How to Determine the Chemical Shifts of Protons Directly Coupled to a Heteronucleus

A very useful form of two-dimensional NMR involves correlation of the chemical shifts of directly coupled nuclei (Freeman and Morris, 1979a; Maudsley and Ernst, 1977), which was pioneered by Ernst and co-workers (1977). The method has two main attractions. The first is the spreading of data in two frequency dimensions. For example, in the case of a ^{1}H–^{11}B correlation the two frequency axes typically have ranges of 40 ppm. This amounts to a tremendous gain in resolution because signals have two frequency coordinates rather than one. The other attraction is that the information obtained can ease the making of assignments. If either partner signal can be assigned, then the correlation gives the assignment of the other.

The experiment, as shown in Fig. 7, consists of two nonselective pulses applied to the S spins, which are separated by the evolution time t_1. The second pulse is applied concurrently with the I spin pulse. The I signal is acquired and Fourier-transformed. The resulting I spin spectrum has the normal frequencies but unusual intensities. The intensities depend on the S chemical shifts and the length of the evolution time t_1.

The origin of the modulation of the intensities of the I signals can be illustrated by an IS spin system with the S and I nuclei coupled only to

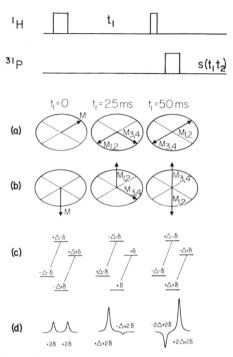

Fig. 7 At the top is the pulse sequence used for correlation of the chemical shifts of I, ^{31}P, and S, ^{1}H, spins. The two proton 90° pulses are separated by the evolution time t_1. The experiment is performed for many regular increments of t_1, and the I spin signal is the one acquired. The signal is a function of t_1 as well as the acquisition time t_2. (a) The S spin magnetization vectors at the end of the evolution time. The $M_{1,2}$ magnetization corresponds to the magnetization between levels 1 and 2. (b) The orientations of the S spin magnetization vectors after the second S pulse applied at the end of the evolution time. (c) The nonequilibrium populations generated by the second S pulse give rise to the populations of the energy levels shown. (d) The intensities of the I signals that arise from the population differences generated by the S pulses. The frequencies of the signals are the same as in a conventional experiment, but the intensities depend on the population differences. The modulation of the intensities of the S spin signals are Fourier-transformed with respect to t_1 to arrive at the two-dimensional spectra.

one another. The energy levels and transitions are shown in Fig. 2. The first S pulse generates transverse S magnetization vectors which freely precess at $\nu_{1,2}$ and $\nu_{3,4}$ during the evolution time, as shown in Fig. 7. The difference in the precessional frequencies is simply J_{IS}, the heteronuclear coupling constant. Depending on the length of the evolution time the S magnetization vectors have different phase angles, as shown in Fig. 7. The second S pulse induces all the S magnetizations to be rotated by 90°. Depending on the phase angle at the end of the evolution time different final magnetizations are obtained. The magnetization may be returned to its equilibrium value $M_{1,2}$ after 25 ms or $M_{3,4}$ after 50 ms, or be inverted, $M_{1,2}$ after 50 ms.

The magnetizations of the S spins correspond to population differences between the energy levels connected by S transitions. The populations generated by the second S pulse are shown in Fig. 7. The population differences are dependent on both the S chemical shifts and the length of the evolution time and can be monitored by a normal I pulse. The resulting I spectra are shown in Fig. 7. The intensities of the I signals deviate from the equilibrium intensities, and the modulation frequencies of the intensities are just the S chemical shifts.

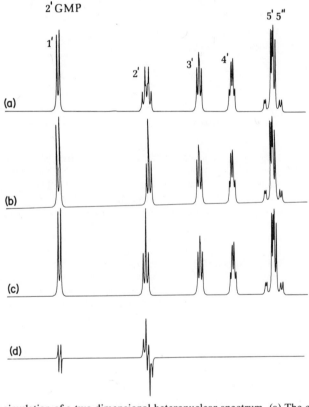

Fig. 8 The simulation of a two-dimensional heteronuclear spectrum. (a) The conventional proton spectrum which is composed of the sum of two subspectra (b and c). Each subspectrum corresponds to the proton spectrum arising from one of the two possible polarizations of the ^{31}P. The heteronuclear two-dimensional spectrum (d) arises from the difference between the two subspectra. The spectra are simulated with line widths of 1 Hz and for a field strength of 200 MHz for protons.

Examination of the mechanism of magnetization transfer points out a key feature of the experiment, which is that the S spectrum obtained in the two-dimensional experiment may be considered to arise from the difference between two subspectra (Bolton and Bodenhausen, 1979). The normal S spectrum may be thought of as arising from the sum of two subspectra—one for each polarization of the I nucleus—whereas the two-dimensional spectrum arises from the difference between the subspectra. This implies that the two-dimensional spectra can be easily interpreted and simulated (Bolton, 1981a,b; Bolton and Bodenhausen, 1979). The simulation is performed by subtracting the S spectrum calculated for the other polarization. An example of this method of simulation is shown in Fig. 8, and the simulation is compared with the experiment in Fig. 9. The straightforward approach just described is applicable to complicated spin systems as long as the proton pulses used are carefully calibrated to 90° (Bolton, 1981a,b).

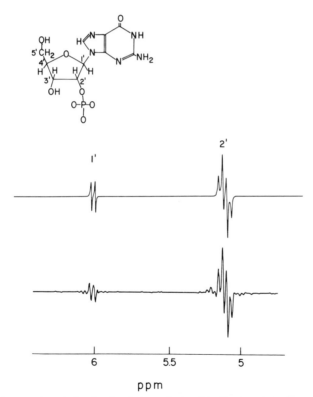

Fig. 9   The top spectrum is for the simulation calculated in Fig. 8, and the bottom spectrum is for the experimental data. For more details see Bolton (1981a,b) and Bolton and Bodenhausen (1979).

In some instances it is advantageous to suppress the heteronuclear coupling in either one or both dimensions. The magnitude of the coupling typically contains little information, and its presence serves only to reduce the signal/noise ratio and obscure the simplicity of the data.

Suppression of the I coupling of the S nuclei may be done in one of the two ways shown in Fig. 10. The use of an I spin 180° pulse to refocus the heteronuclear coupling is commonly employed because of its experimental simplicity (Freeman and Morris, 1979a; Maudsley and Ernst, 1977). However, this method can give rise to artifacts if there are two S signals, whose I spin-decoupled chemical shifts are separated by $^1J_{IS}/2$, that are coupled to one another (Bolton, 1983a). The use of "true decoupling" is the preferred approach and eliminates the anomalous signals that occasionally occur when refocusing is used. At the end of t_1 the two S magnetization vectors of the IS spin system are in phase with one another. A delay time Δ_H is used to allow the two S magnetization vectors to become

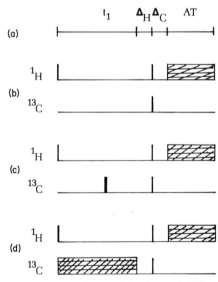

Fig. 10 Pulse sequences used for correlation of the chemical shifts of ^{13}C and ^1H using $^1J_{CH}$ as the linkage for magnetization transfer. The timing of all the pulse sequences is given in (a), with t_1 being the evolution period and Δ_H and Δ_C being delay times which allow the magnetization vectors to have the proper phases. During Δ_H the proton magnetization vectors become out of phase with respect to $^1J_{CH}$, and during Δ_C the carbon magnetizations become in phase with respect to $^1J_{CH}$. In pulse sequence (b) the proton signals obtained are coupled to ^{13}C, but the ^{13}C signals are decoupled from protons. In pulse sequence (c) a ^{13}C 180° pulse is applied at the middle of t_1 to suppress the ^{13}C coupling of the protons. ^{13}C decoupling is used in pulse sequence (d) to suppress the ^{13}C coupling of the protons. The cross-hatched areas indicate decoupling.

180° out of phase with one another and to permit efficient magnetization transfer which is solely dependent on the I spin decoupled S chemical shift. The magnetization transfer is completed by the I pulse at the end of $t_1 + \Delta_H$. Turning on S decoupling at this time eliminates the magnetization transfer. A delay time Δ_C is used as in INEPTR to allow the I magnetization vectors, which were originally of opposite algebraic sign and hence phase, to come into phase with one another. As was the case with the INEPTR experiments, a compromise choice for Δ_C is used.

An example of cross-correlation is shown in Fig. 11. Each ^{13}C proton pair gives rise to a signal that has two coordinates—one for the chemical shift of each nucleus. For this reason the spectra are often referred to as maps.

Two-dimensional correlation of the chemical shifts of nuclei has been performed using ^{13}C (Freeman and Morris, 1979a; Maudsley and Ernst, 1977), ^{15}N (Gray), ^{11}B (Finster *et al.*, 1980), and ^{31}P (Bolton and Bodenhausen, 1979; Bolton, 1981a,b) as the I nuclei and ^{1}H as the S nuclei. There is no reason why the method cannot be applied to many other pairs of nuclei. One promising application is to use known I or S assignments to

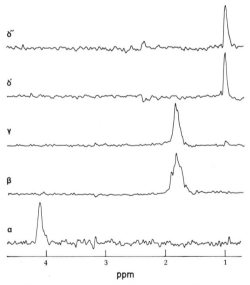

Fig. 11 The spectra of leucine are shown in the phase-sensitive mode of display and have been extracted from the complete correlation map. The spectra correspond to the proton shifts along the horizontal, and the five traces shown are for the chemical shifts of the five nonequivalent ^{13}C chemical shifts. The spectra show that the chemical shifts of the protons can be determined via ^{13}C detection, and the spectra give the correlation between the ^{1}H and ^{13}C chemical shifts.

determine the assignments of the correlated nucleus whose chemical shifts are not as well understood.

The two-dimensional experiments previously described may be thought of as a generalization of internuclear double resonance (INDOR). This method is nearly as old as magnetic resonance itself. For example, in 1954 Royden determined the gyromagnetic ratio of ^{13}C in solution by observing the collapse of the ^{1}H multiplet of enriched methyl iodide when a strong rf field was applied near the resonant frequency of the ^{13}C nuclei (Royden, 1954). This technique was thoroughly examined by Anderson, Freeman, and others who developed a wide array of INDOR experiments for detecting the signs of coupling constants and multiple-quantum effects among other effects (Anderson and Freeman, 1963; Freeman, 1965). If one is interested in determining the chemical shift of a single S nucleus coupled to an I nucleus, time can be saved by using Royden's approach. However, the two-dimensional approach has the advantage that the experiment works whether or not the chemical shift of the S nuclei can even be estimated, whereas the INDOR experiment typically requires at least a good estimate of the chemical shift for the experiment to be performed in a reasonable amount of time.

VII. How to Determine Whether a Heteronucleus Is Directly Coupled to a Like Heteronucleus

If the heteronucleus occurs in high natural abundance, greater than about 15–20%, or has been enriched to this level, then the problem can be rather easily solved. One can chose between homonuclear decoupling and homonuclear J spectroscopy. The homonuclear chemical shift correlation experiment is a two-dimensional one, allows simultaneous determination of the chemical shifts of scalar-coupled pairs of homonuclei (Wider *et al.*, 1981; Nagayama *et al.*, 1979), and has been applied to ^{11}B (Venable *et al.*, 1982). The homonuclear J experiment is also performed in a manner analogous to that used for the heteronuclear case, and heteronuclear and homonuclear couplings are easily distinguished one from the other (Hall and Sukumar, 1979).

The case that is more challenging occurs when the natural abundance of the heteronucleus is about 1%. In this situation the previously described experiments do not succeed, because the signals from heteronuclei with magnetically inactive neighbors swamp out those of heteronuclei with magnetically active neighbors. One approach to the selective detection of only nuclei that are coupled to like nuclei is multiple-quantum spectroscopy. This technique has the basic feature of detecting only nuclei coupled to like nuclei.

This approach was used by Freeman and associates for the detection of ^{13}C couplings in natural abundance (Bax *et al.*, 1980a,b). Because the natural abundance of ^{13}C is 1.1%, only one site in 10^4 is ^{13}C and has one ^{13}C neighbor. The experiment relies on observing only the signals from ^{13}C sites that have ^{13}C neighbors. The selectivity arises from detecting only signals that once existed as carbon–carbon double-quantum coherence. This experiment, known as INADEQUATE, gives a one-dimensional spectrum which is the ^{13}C-coupled ^{13}C spectrum.

The basic INADEQUATE experiment has been superseded by correlation of the chemical shifts of adjacent carbon sites in a molecule (Bax *et al.*, 1980a,b). The basic idea is the same: detection of only ^{13}C sites within ^{13}C neighbors. However, the results are easier to interpret because the map has signals whose coordinates are the chemical shifts of the adjacent sites. The method is an improvement over INADEQUATE because the relatively large $^1J_{CC}$, 30–70 Hz, can cause overlap in the one-dimensional spectrum, hence the higher resolving power of the two-dimensional approach is needed.

The two-dimensional experiment is known as a carbon–carbon connectivity plot (CCCP). The results shown in Fig. 12 allow determination of all the carbon–carbon linkages (Patt *et al.*, 1982). The resolution needed to determine the linkages is not demanding, and as few as 64 increments of the evolution time can be sufficient (Patt *et al.*, 1982). This approach has also been used for selective detection of protons coupled to at least one other proton (Bodenhausen and Dobson, 1981; Piantini *et al.*, 1983). The experiment can also be applied, for example, to ^{113}Cd-metalothionein complexes and ^{11}B.

The most serious drawback in determining the presence of linkages in this manner is the very low sensitivity. This very low sensitivity implies that large amounts, 0.5 g or more, of sample are required with overnight runs for ^{13}C. However, nuclei of higher natural abundance such as ^{113}Cd, ^{11}B, and ^{31}P should be quite suitable.

VIII. How to Determine Whether Two Nuclei Belong to the Same Coupling Network

In many cases it is of interest to determine extended coupling networks. The heteronuclear chemical shift correlation experiment determines coupling networks that contain only two nuclei. Consider the case shown

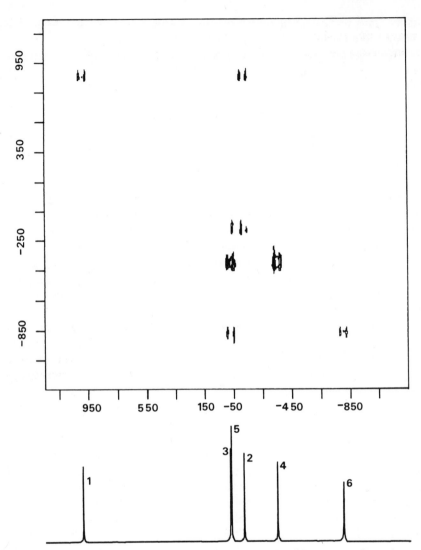

Fig. 12 The bottom spectrum is the normal proton-decoupled ^{13}C spectrum of α-D-glucose, and the signals have been assigned as shown. The ^{13}C spectrum is aligned with the CCCP map above it. The horizontal axis of the CCCP map is the ^{13}C chemical shifts, and the frequencies relative to the carbon transmitter are indicated. The vertical axis is the double-quantum axis. The horizontal and vertical components of a signal indicate which two carbon sites are adjacent. For example, since C-1 and C-2 are adjacent, a set of signals is present whose horizontal component is the chemical shift of C-1 and whose vertical component is the sum of the chemical shifts (in frequency relative to the transmitter frequency) of C-1 and C-2. The signal is a doublet because it is split by $^{1}J_{CC}$. A complementary signal appears whose horizontal component is the chemical shift of C-2 and whose vertical compo-

here, in which one proton is directly coupled to a heteronucleus. The directly coupled proton is referred to as the neighbor (M) and the heteronucleus as the spy (X). The extended coupling network also includes protons coupled to the neighbors. These more distant protons are called remote (A). For A—M—X, $J_{AM} \neq 0$, $J_{AX} = 0$, and $J_{MX} \neq 0$.

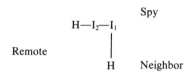

One kind of information that can be extracted from extended coupling networks is the determination of which boron sites are adjacent to one another. For example, two adjacent boron sites share neighbor and remote protons, and their extended coupling networks overlap.

Another situation in which the determination of extended coupling networks is of value is investigation of the chemical shifts and scalar couplings of protons that are anywhere near the heteronucleus used as the spy. Because the spy reports on both the neighbor protons as well as the remote protons, considerably more information can be gained than in a conventional chemical shift correlation experiment. The method can be extended to spying on protons even more distant than the remote protons.

The experimental procedure involves the merging of two distinct magnetization transfer processes (Bolton, 1982; Bolton and Bodenhausen, 1982; Eich *et al.,* 1982). The first transfer is between protons and allows information about the remote protons to be transferred to the neighbors. The second step is a heteronuclear magnetization transfer which transfers information about the remote and neighbor protons to the heteronucleus. The experiment is the merging of a homonuclear proton–proton chemical shift correlation with a heteronuclear chemical shift correlation.

The actual experiment can be most easily illustrated by considering an A—M—X spin system. It is assumed that only the M (neighbor) spin is coupled to the heteronucleus X (spy) and that there is coupling between the A (remote) and neighbor nuclei. Figure 13 shows a pulse sequence used for a two-step magnetization transfer.

The first part of the pulse sequence involves the transfer of magnetization from remote to neighbor spins in a manner directly analogous to a

---

nent is the sum of the chemical shifts of C-1 and C-2. The map contains sufficient information to assign all the ^{13}C signals of this sample. The data were obtained by S. L. Patt who has discussed this and other examples in detail (Patt *et al.,* 1982).

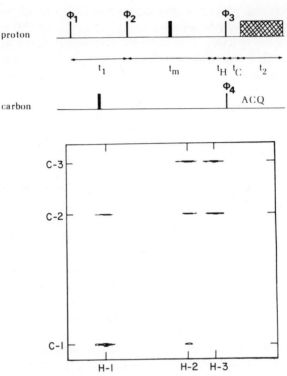

Fig. 13 At the top is the pulse sequence used for ¹³C detection of remote protons via relayed coherence transfer spectroscopy. The new feature of the experiment is the inclusion of a mixing time which has a proton 180° pulse in the middle. The mixing time together with the final proton pulse allows the transfer of magnetization from a remote to a neighbor proton. The experiment is described elsewhere in more detail (Bolton, 1982; Bolton and Bodenhausen, 1982). The contour map shows that data for 1-propanol. The data obtained by the relayed transfer method correlate the chemical shifts of ¹³C nuclei not only with protons directly bonded but to protons coupled to the directly coupled protons. Thus sites C-1 and C-2 must be adjacent because they correlate with two common proton frequencies. Similarly C-2 and C-3 are adjacent. The contour map readily allows determination of the structure. The proton signals can be assigned readily as shown. For example, H-1 is assigned because its signal correlates with C-1 and C-2 but not C-3. The relayed transfer map gives the structure and all the ¹H and ¹³C assignments directly. Application to more complicated cases is presented in Bolton (1982).

proton–proton correlation (Hall and Sukumar, 1979; Wider *et al.*, 1981). The first two proton pulses, separated by the evolution time t_1, transfer magnetization from the A to the M spins with an efficiency that depends on the chemical shift of the A spins and the duration of t_1, as well as the strength of the AM coupling. The transfer of magnetization from the M to the X spins is accomplished by the simultaneous proton and spy nucleus pulses at the end of the time $t_m + t_H$. The time $t_m + t_H$ is chosen such that the multiplets that are out of phase with respect to J_{AM} at the beginning of this time become in phase and the multiplets that are in phase with respect to J_{AX} become out of phase (Bolton, 1982; Bolton and Bodenhausen, 1982).

Because J_{AX} is much larger than J_{AM} a proton decoupling pulse is applied at the middle of t_m so that at the end of t_m the multiplets are in phase with respect to J_{MX} but out of phase with respect to J_{AX} at the end of $t_m + t_H$. The time $t_m + t_H$ is chosen to be about $\frac{1}{4}J_{AM}$ for a typical $^3J_{HH}$, so that there is no strong selection in favor of a particular proton multiplicity.

The pulse sequence shown in Fig. 13 also correlates the chemical shifts of the neighbor and spy nuclei. Therefore a spy nucleus decoupling pulse is applied at the middle of t_1, suppressing the heteronuclear coupling of neighbor protons. To enhance both sensitivity and simplicity a delay t_C is used after the final magnetization transfer to allow the transferred magnetization to be detected with proton decoupling. The t_C delay is analogous to that used in INEPTR and heteronuclear chemical shift correlation spectroscopy. The net result of the pulse sequence is correlation of the heteronuclear decoupled remote and neighbor proton signals with the spy signal.

A simple example is shown in Fig. 13 which displays the related transfer data for 1-propanol as a contour map. In general, sites at the end of a chain correlate with two proton signals, whereas those in the interior correlate with three signals. The data for 1-propanol is particularly easy to interpret and directly gives the molecular structure and the assignments of all the ^1H and ^{13}C signals. The data show that C-1 and C-2 are adjacent, because these carbon sites correlate with two common proton frequencies. Similarly, C-2 is adjacent to C-3. The proton assignments can also be made. H-1 is assigned as shown because H-1 correlates with C-1 and C-2 but not C-3. The assignment of H-3 is based on the correlation of this signal with C-2 and C-3 but not C-1. The assignment of H-2 is based on the correlation of this signal with all three carbon sites.

Relayed magnetization transfer experiments can also be used to investigate molecular structures using a heteronucleus as a spy. Figure 14 shows the data obtained using ^{31}P detection of the protons of phosphothreonine. The relayed transfer data contain signals from all the nonexchangeable protons. The spectrum can be easily analyzed to give the coupling constants and chemical shifts of the protons. The relayed transfer spectrum may be thought of as the sum of the heteronuclear and homonuclear chemical shift correlation experiments. The heteronuclear contribution is the difference between the two subspectra associated with the two polarizations of the ^{31}P nucleus. The homonuclear contribution is the differ-

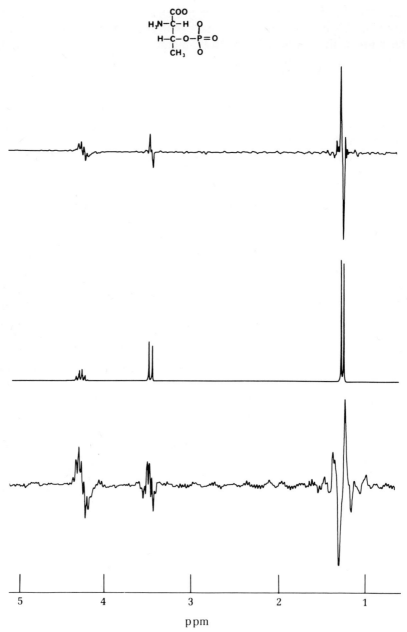

Fig. 14 The middle spectrum is a simulation of the normal proton spectrum of phosphothreonine. The top spectrum is a relayed coherence transfer spectrum obtained using a constant mixing time. The relayed transfer spectrum can be thought of as the sum of the

ence between the two subspectra associated with the two polarizations of the neighbor proton and contains the signals of the remote protons. The sum of these two contributions is the relayed transfer spectrum (Bolton and Bodenhausen, 1982).

The ^{13}C detection experiment offers detailed structural information with a sensitivity not much less than that of a conventional heteronuclear chemical shift correlation experiment. The only experiments that offer comparable information are those involving INADEQUATE and CCCP which can use only 1% of the ^{13}C sites. Because the relayed experiment uses all the ^{13}C sites, its sensitivity is much greater than that of the INADEQUATE and CCCP experiments.

IX. Resolution Enhancement Techniques

Over the years a variety of methods have been proposed for enhancing the resolution of spectra. These methods have ranged from the elimination of broad components by data manipulation (Ferrige and Lindon, 1980), restricting the volume of the sample observed by pulsed field inhomogeneity gradients (Bax and Freeman, 1980), selective excitation of a narrow region of a line (Morris and Freeman, 1978) as in a "hole-burning" microwave or optical experiment, and exploiting the relatively rapid T_2 relaxation of broad components to arrive at a sharp final spectrum (Campbell, 1979).

All the conventional methods for enhancing resolution are based on discriminating against part of the sample on the basis of spatial location or relaxation rate. Thus resolution enhancement always requires greater sensitivity than a normal experiment. A variety of resolution enhancement techniques not involving two-dimensional spectroscopy have been reviewed by Ferrige and Lindon (1980). To a large extent all these methods have been superseded by two-dimensional experiments. Two-dimensional experiments offer a variety of ways to overcome spectral overlap.

---

normal heteronuclear two-dimensional spectrum and the homonuclear two-dimensional spectrum. The homonuclear spectrum, in the limit of weak coupling, is the difference between the subspectra corresponding to the two polarizations of the neighbor proton. The bottom spectrum shows the data obtained for a variable mixing time. The variable mixing time does not give rise to data amenable to extraction of coupling constants that can be related to conformational features. The data obtained with a constant mixing time can be readily analyzed to give both chemical shifts and couplings. It is noted that the relayed transfer data for phosphothreonine allow determination of the chemical shifts and coupling constants of all the nonexchangeable protons of phosphothreonine via ^{31}P detection.

One very promising approach to resolution enhancement is the use of multiple-quantum NMR (Bodenhausen, 1981). At present there are only a few examples of this potentially useful approach. The utility of the multiple-quantum experiment is that, although both chemical shifts and splittings due to spin–spin couplings scale with the number of quanta involved, the line widths can remain constant, or even decrease, with the number of quanta. This approach is especially appealing in applications to quadrupolar nuclei which tend to have rather broad lines. The multiple-quantum technique has been applied to ^2H and ^{14}N to obtain high-resolution spectra (Prestegard and Miner, 1981).

X. Concluding Remarks

The preceding discussion describes a number of applications of relatively new developments in NMR. The ability to transfer magnetization from nucleus to nucleus not only allows sensitivity enhancement but also provides a rich source of information that can aid structural and conformational studies. Although some of the two-dimensional experiments may seem to be somewhat esoteric, it is most likely that in a relatively short time many of these procedures will become rather mundane and applied to a much wider variety of nuclei then currently described in the literature.

Note Added in Proof: The methods for enhancing the sensitivity of low-gyromagnetic ratio nuclei by magnetization transfer and the one-dimensional method of determining the number of protons coupled to a heteronucleus have been improved and evaluated by Sørenson and Ernst (1983). An advance in the two-dimensional method of determining heteronuclear couplings that allows selective removal of either long-range or short-range couplings has been presented by Bax (1983). The methodology of the relayed transfer experiment has been expanded (Bolton, 1983b; Kogler *et al.,* 1983).

Acknowledgments

The author would like to thank Steven L. Patt for the data in Fig. 12. The author's research described in this chapter was supported at various times by the National Science Foundation and National Institutes of Health of the United States, as well as by the Petroleum Research Fund of the American Chemical Society and the Camille and Henry Dreyfus Foundation.

References

Allerhand, A., Doddrell, D., and Glushko, V. (1972). *J. Chem. Phys.* **56**, 3683.
Anderson, W. A., and Freeman, R. (1962). *J. Chem. Phys.* **37**, 85.

Anderson, W. A., and Freeman, R. (1963). *J. Chem. Phys.* **39**, 806.

Bax, A. (1983). *J. Magn. Reson.* **52**, 330.

Bax, A., and Freeman, R. (1980). *J. Magn. Reson.* **37**, 177.

Bax, A., Freeman, R., and Kempsell, S. P. (1980a). *J. Am. Chem. Soc.* **102**, 4849.

Bax, A., Freeman, R., and Kempsell, S. P. (1980b). *J. Magn. Reson.* **41**, 349.

Bodenhausen, G. (1981). *Prog. NMR Spectrosc.* **14**, 137.

Bodenhausen, G., and Dobson, C. M. (1981). *J. Magn. Reson.* **44**, 212.

Bodenhausen, G., and Roberts, M. F. (1980). *FEBS Lett.* **117**, 311.

Bodenhausen, G., and Ruben, D. J. (1980). *Chem. Phys. Lett.* **69**, 185.

Bolton, P. H. (1980). *J. Magn. Reson.* **41**, 287.

Bolton, P. H. (1981a). *In* "Biomolecular Stereodynamics" (R. H. Sarma, ed.), Vol. 2, p. 437. Adenine, New York.

Bolton, P. H. (1981b). *J. Magn. Reson.* **45**, 239.

Bolton, P. H. (1982). *J. Magn. Reson.* **48**, 336.

Bolton, P. H. (1983a). *J. Magn. Reson.* **51**, 134.

Bolton, P. H. (1983b). *J. Magn. Reson.* (In press.)

Bolton, P. H., and Bodenhausen, G. (1979). *J. Am. Chem. Soc.* **101**, 1080.

Bolton, P. H., and Bodenhausen, G. (1982). *Chem. Phys. Lett.* **89**, 139.

Bolton, P. H., and James, T. L. (1980). *Biochemistry* **19**, 1388.

Brevard, C., and Schimpf, R. (1982). *J. Magn. Reson.* **47**, 528.

Burum, D. P., and Ernst, R. R. (1980). *J. Magn. Reson.* **39**, 163.

Campbell, I. D. (1979). *Methods Biochem. Anal.* **25**, 1.

Chingas, G. C., Garroway, A. N., Moniz, W. B., and Bertrand, R. D. (1979). *J. Magn. Reson.* **35**, 283.

Chingas, G. C., Garroway, A. N., Moniz, W. B., and Bertrand, R. D. (1980). *J. Am. Chem. Soc.* **102**, 2526.

Doddrell, D. M., Pegg, D. T., and Bendall, M. R. (1982). *J. Magn. Reson.* **48**, 323.

Eich, G., Bodenhausen, G., and Ernst, R. R. (1982). *J. Am. Chem. Soc.* **104**, 3731.

Ernst, R. R. (1966). *Adv. Magn. Reson.* **2**, 1.

Ferrige, A. G., and Lindon, J. C. (1980). *Prog. NMR Spectrosc.* **14**, 27.

Finster, D. C., Hutton, W. C., and Grimes, R. N. (1980). *J. Am. Chem. Soc.* **102**, 400.

Freeman, R. (1965). *J. Chem. Phys.* **43**, 3087.

Freeman, R., and Morris, G. A. (1978). *J. Magn. Reson.* **29**, 433.

Freeman, R., and Morris, G. A. (1979a). *Bull. Magn. Reson.* **1**, 5.

Freeman, R., and Morris, G. A. (1979b). *J. Am. Chem. Soc.* **101**, 760.

Freeman, R., Morris, G. A., and Turner, D. L. (1977). *J. Magn. Reson.* **26**, 373.

Freeman, R., Grutzner, J. B., Morris, G. A., and Turner, D. L. (1978). *J. Am. Chem. Soc.* **100**, 5637.

Freeman, R., Mareci, T. H., and Morris, G. A. (1981). *J. Magn. Reson.* **42**, 341.

Hall, L. D., and Sukumar, S. (1979). *J. Am. Chem. Soc.* **101**, 3120.

Jakobsen, H. J., Sørenson, O. W., Brey, W. S., and Kanyha, P. (1982). *J. Magn. Reson.* **48**, 328.

Kogler, H., Sørenson, O. W., Bodenhausen, G., and Ernst, R. R. (1983). *Chem. Phys. Lett.* (In press.)

Maudsley, A. A., and Ernst, R. R. (1977). *Chem. Phys. Lett.* **50**, 368.

Morris, G. A. (1980). *J. Am. Chem. Soc.* **102**, 428.

Morris, G. A., and Freeman, R. (1978). *J. Magn. Reson.* **29**, 433.

Müller, L., Kumar, Anil, and Ernst, R. R. (1977). *J. Magn. Reson.* **25**, 383.

Müller, L. (1979). *J. Magn. Reson.* **36**, 301.

Nagayama, K, Wütrich, K., and Ernst, R. R. (1979). *Biochem. Biophys. Res. Commun.* **90**, 305.

Noggle, J. H., and Shirmer, R. E. (1971). "The Nuclear Overhauser Effect: Chemical Applications," Academic Press, New York.

Patt, S. L., Sauriol, F., and Perlin, A. S. (1982). *Carbohy. Res.* **107,** C1.

Patt, S. L., and Shoolery, J. N. (1982). *J. Magn. Reson.* **46,** 535.

Piantini, U., Sørenson, O. W., and Ernst, R. R. (1983). *J. Am. Chem. Soc.* **104,** 6800.

Prestegard, J. H., and Miner, V. W. (1981). *J. Am. Chem. Soc.* **103,** 5979.

Royden, V. (1954). *Phys. Rev.* **96,** 543.

Solomon, I. (1955). *Phys. Rev.* **99,** 559.

Sørenson, O. W., and Ernst, R. R. (1983). *J. Magn. Reson.* **51,** 477.

Venable, T. L., Hutton, W. C., and Grimes, R. N. (1982). *J. Am. Chem. Soc.* **104,** 4716.

Vidusek, D. A., Roberts, M. F., and Bodenhausen, G. (1982). *J. Am. Chem. Soc.* **104,** 5452.

Waugh, J. S., Huber, L. M., and Haeberlen, U. (1968). *Phys. Rev. Lett.* **20,** 180.

Wider, G., Baumann, R., Nagayama, K., Ernst, R. R., and Wütrich, K. (1981). *J. Magn. Reson.* **42,** 73.

3 Computational Considerations

Charles L. Dumoulin
George C. Levy

Department of Chemistry
Syracuse University
Syracuse, New York

I. Data Acquisition

A. Basics

It is useful to review certain basic concepts important in the study of exotic nuclei NMR. Nuclear magnetic resonance experiments with these

NMR OF NEWLY ACCESSIBLE NUCLEI, VOL. 1

nuclei are greatly affected by the nature of the nuclei themselves. Thus experimental and instrumentation requirements vary from nucleus to nucleus. Often these requirements are much more strict than those found in ^{1}H and ^{13}C NMR.

Perhaps the most serious experimental restriction imposed by nature is the large chemical shift range exhibited by many exotic nuclei (Harris and Mann, 1978). The Nyquist theorem dictates that, to measure a sine wave unambiguously, the signal must be sampled (digitized) at least two times per cycle. This means that for a spectral width of 100 kHz the data must be digitized and recorded at a rate of 200 kHz. If quadrature detection is used, data can be simultaneously acquired on two channels, allowing negative frequencies to be distinguished from positive ones. Consequently, for a given spectral window, a twofold reduction in the sampling frequency is possible.

Another problem associated with wide chemical shift ranges is the irradiation spectrum afforded by the excitation pulse. The size of the irradiation spectrum of a rectangular pulse is always inversely proportional to the pulse length. Unfortunately, for a given hardware configuration (pulse amplifier, probe, etc.), a decrease in the excitation pulse length results in a decrease in the tip angle. The shape and amplitude of the excitation spectrum of a pulse are determined only by the length of the pulse and the strength of the rf field. The tip angle applied to a spin, on the other hand, is determined by the rf field applied to the spin and the duration of the pulse. Consequently, increasing a pulse length increases all spin tip angles only as long as the pulse excitation spectrum remains broad enough to cover the spins of interest.

Often the dominant problem in NMR experiments with exotic nuclei is low sensitivity. Much progress has been made in improving the sensitivity of ^{13}C-NMR experiments in recent years, and several of the techniques that have been developed can be applied to other nuclei. These techniques include the following:

(a)   signal averaging in the time domain followed by Fourier transformation (FT) (1960s),

(b)   nuclear Overhauser enhancement (1960s),

(c)   better probe designs (Hoult and Richards, 1976) (mid-1970s to the present),

(d)   multipulse magnetization transfer techniques: INEPT (Morris and Freeman, 1979), cross-polarization (Pines *et al.*, 1973), etc.

Nevertheless, careful attention must be paid to all experimental details when spectra with inherently low signal/noise ratios are acquired.

B. Computer Requirements for a Multinuclear NMR Spectrometer

The proper selection of a computer and its configuration are extremely important when building or purchasing a multinuclear NMR spectrometer. It is often difficult to reach a compromise between anticipated computer needs and budgetary limitations. Requirements for memory, disk storage, graphic display, plotting, analog-to-digital conversion rates, spectrometer interface, and pulse programming must be carefully evaluated. An important consideration frequently underestimated is the quality and quantity of available software. The computer is so central to experiment control and the user–spectrometer interface that a poor choice of computer configuration can greatly reduce the throughput and efficacy of a spectrometer.

If nuclei with large chemical shift ranges are of interest to the spectroscopist, high-speed analog-to-digital conversion is important, if not essential. Two digitizing channels allow the full advantage of quadrature detection to be realized. Each channel should have programmable filters to minimize the aliasing of noise into spectra and to sample and hold circuitry to ensure accurate digitization. Another desirable feature is computer compensation for any dc offsets in the signal. This feature, however, is less important when a full-phase alternation scheme such as CYCLOPS (Hoult and Richards, 1974) is used.

The digitizer resolution should also be under computer control so that the data can be efficiently digitized. For example, at the beginning of a long experiment, the number of digitizer bits can be set to maximum (typically 12 or 13). When an overflow of data is imminent, the data can be scaled down and the digitizer resolution adjusted to a lower setting to match the scaling of the data. Data acquisition can then continue until the programmed number of free induction decays (FIDs) have been acquired or the digitizer can no longer be scaled back. For a spectroscopist concerned with spectra exhibiting a high dynamic range, proper use of digitizer resolution is particularly important.

A computer feature that has become more common is control of spectrometer functions. Commercial vendors are beginning to offer spectrometers controlled *entirely* through the computer keyboard. Modern NMR system computers often control sample temperature, spectrometer rf's and magnet shimming, in addition to all data acquisition parameters. Even automated sample changing has been demonstrated. Computer control of a spectrometer allows the spectroscopist to perform complex experiments and/or multiple experiments without operator intervention.

New experimental methods such as INEPT, multiquantum NMR, two-dimensional Fourier transformation (FT) NMR, and selected excitation require accurate computer control of pulse timing, phase, and amplitude. Many experiments also require full pulse control of both the observe and decouple channels. Pulse sequences can be extremely complex. Consequently, provision for *flexible* pulse programming by users is a necessity in modern spectrometer systems.

Advances in computer hardware and software have greatly expanded the flexibility and capability of NMR spectrometers. Significant increases in sample throughput and spectrometer usage have been obtained by designs that allow simultaneous data acquisition and data processing. Two design philosophies exist: the use of a single multitask computer for all functions and the use of two or more computers, each dedicated to specific tasks.

As hardware costs have decreased, newer spectrometer designs have begun to incorporate multiple computers. Laboratorywide computer networks (Levy, *et al.*, 1981) have also been demonstrated but are practical only for larger laboratories having several spectrometers (not all of which have to be NMR spectrometers).

II. Basic Data Reduction

A. Preliminary Processing

After the data are acquired, the computer stores (either on a disk or in memory) a series of numbers that represent the digitized time response of the NMR spin system to the excitation pulse. The digitized time response or FID must first be normalized so that any dc offsets in the data are removed. The size of the data set (i.e., the number of data points) can then be increased by appending zeros to the FID. Next, weighting functions can be applied to enhance the signal/noise ratio or the resolution. A fast Fourier transformation (FFT) is then performed to convert the data from the time domain to the frequency domain.

1. Baseline Normalization

If offsets in a FID are not removed, troublesome zero-frequency spikes can appear in the frequency domain spectrum. Phase alternation schemes during data acquisition are very effective in canceling offsets. If phase alternation is not used, however, a simple computer algorithm can be employed. This algorithm simply averages data from the last part of the

FID and subtracts this average number from the entire FID. Real and imaginary components are independently corrected if the data are complex. Typically, the last 10% of the FID is used to ensure that low-signal-frequency components do not contribute to calculation of the FID baseline. This algorithm is outlined in Eqs. (1) and (2) for a FID of length N, where data$_{corr}$ represents the corrected data. Only data points between data(M) and data(N) contribute to calculation of the offset:

$$\text{data}_{corr}(i) = \text{data}(i) + \text{offset} \tag{1}$$

where

$$\text{offset} = \frac{N - M}{N} \sum_{i=M}^{N} \text{data}(i) \tag{2}$$

2. Zero-Filling

Zero-filling is a convenient method for increasing the digital resolution of a spectrum without increasing the experiment time. Zero-filling simply increases the length of a FID by appending zeros. Since the spectral window (or frequency range) of the spectrum is determined only by the time interval between successive digitizations, appending zeros increases the number of data points without increasing the spectral width. Thus each spectral peak obtained after FT consists of more points. Care must be taken to ensure that the signals present in the FID decay completely before zeros are appended. Otherwise, negative lobes due to truncation errors will appear at the base of spectral peaks. Zero-filling is useful mainly for better quantification of peak intensities and integrals. Zero-filling a FID more than once does not increase the information content but *can provide* better defined peaks. This is equivalent to performing a mathematical interpolation between data points in the final frequency spectrum.

3. Weighting Functions

Perhaps one of the most useful properties of the FT is the ability to convolute two functions easily. Convoluting two functions in one Fourier domain is mathematically equivalent to multiplying the two functions in the other domain. Convolutions are typically used to modify resolution or the signal/noise ratio. In general, a trade-off exists, and increasing the resolution decreases the signal/noise ratio, and vice versa. Several weighting functions are generally available to the NMR spectroscopist. Some of these will be outlined briefly here.

(a) *Trapezoidal multiplication*. Trapezoidal multiplication is the simplest weighting function that can be applied to a FID. This weighting function is simply equal to *one* for a specified number of data points, and beyond that point the weighting function decreases linearly until it becomes *zero* at the last data point. Most software systems also allow a linear increase in this weighting function at the beginning of the data set. Thus trapezoidal weighting can simultaneously reduce broad spectral features and increase the signal/noise ratio. Unfortunately, spectral line shapes can become distorted if severe weighting parameters are used. The form of the trapezoidal weighting function is shown in Fig. 1a.

The main advantage of trapezoidal weighting is its programming and operational simplicity. More complex weighting functions are generally superior and are usually implemented.

(b) *Exponential multiplication*. Weighting the experimental data with a decreasing exponential has become the method of choice when the signal/noise ratio is to be maximized. The exponential weighting function has several advantages over other weighting functions. The primary advantage is that exponential weighting of a FID is equivalent to convoluting the spectra with a Lorentzian line shape. Since the theoretical line shape of most NMR signals is also Lorentzian, the shape of the convoluted spectral lines remains Lorentzian. The input parameter for exponential weighting is generally called the line broadening factor. This factor

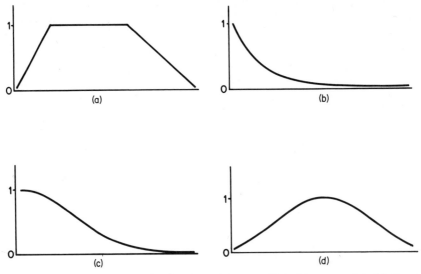

Fig. 1 Commonly used weighting functions: (a) trapezoidal, (b) exponential, (c) Gaussian, (d) shifted Gaussian.

represents the width that will be added to each spectral line. It has been shown (Bartholdi and Ernst, 1973) that a maximum signal/noise ratio is obtained when the line broadening is chosen to be equal to the true experimental line width. The use of an exponential in this manner describes the application of a matched filter.

The form of the exponential weighting function is given in Eqs. (3) and (4) and is shown in Fig. 1b.

$$W_i = \exp(-at_i) \tag{3}$$

where t_i is the time of the ith data point ($0 \leq t_i \leq$ acquisition time), a the line broadening parameter, and W_i the weighting applied to the ith point.

If the line broadening is chosen to be negative, resolution enhancement can be performed at the expense of the signal/noise ratio. Resolution enhancement is better achieved, however, by other weighting functions.

(c) *Gaussian multiplication.* Gaussian weighting functions are extremely useful for resolution enhancement (Ferrige and Lindon, 1978). This is due to several factors: First, the Gaussian line shape has much less signal intensity at the base than the Lorentzian line shape and, second, the FT of a Gaussian line shape is also Gaussian. Typically, the Lorentzian decay of the FID is first canceled by an increasing exponential weighting function. The Gaussian function is then applied to the data. The center of symmetry of the function can be placed anywhere along the data. Placement of the weighting function toward the end of the FID discriminates against wide peaks but also yields low signal/noise ratios. Higher signal/noise ratios are obtained when the weighting function is applied at the beginning of the FID. Unless the Gaussian weighting function is centered at the midpoint of the FID, however, small contributions from dispersive components may distort line shapes because the weighting function is not truly symmetric. Because Gaussian multiplication yields Gaussian line shapes, the magnitude spectrum does not have any contributions from dispersive Lorentzian components and thus the magnitude spectrum (described later) has line shapes similar to those of a phased spectrum. The form of the Gaussian weighting function (including cancelation of exponential decay) is given in Eq. (4) and shown in Fig. 1c.

$$W_i = \exp\{\pi at_i - (t_x - t_i)^2[(3\pi b)^2/25]\} \tag{4}$$

where t_i is the time of acquisition for the ith point, a the Lorentzian line width to be compensated for with an increasing exponential, b the Gaussian line width to be applied to the data, and t_x the time of the center of symmetry of the Gaussian weighting function.

(d) *Resolution enhancement.* Wide spectral components have more of their intensity at the first part of the FID than narrow components. Consequently, a weighting function can discriminate against wide spectral components by minimizing intensity from the first part of a FID. A simple resolution enhancement scheme (DeMarco and Wuthrich, 1976) involves the use of a sinusoidal curve as a weighting function. The period of the sinusoidal curve is chosen to be two times the acquisition time, so that the first and last points of the FID are zeroed. Having the weighting function return to zero at the end of the FID ensures that line shape distortions due to truncation of the FID are minimized. Variations of this method have been investigated (Lindon and Ferrige, 1980), and the utility of phase-shifted sinusoidal weighting functions has been demonstrated.

4. Fourier Transformation

Fourier transformation is unquestionably the most important software tool for pulse NMR data processing. If FT did not exist, high-resolution pulse experiments would be essentially useless. If the FFT algorithm (Cooley and Tukey, 1965) did not exist, the much slower discrete FT algorithm would have to be used. The FFT algorithm is different from other software moieties because of its lack of user interaction. Rarely must the user worry about the effect of word size, quadrature detection, and data set size on the FFT algorithm itself. A clear understanding of these parameters and how they affect data before and after FT, however, is essential.

Fourier transformation can be simplistically viewed as a mathematical mixing of each data point with all other data points. Thus the modification of one point prior to FT affects all points after FT. This is best illustrated by either zeroing or overflowing the first data point of a FID. The transformed data have distorted line shapes. The extent of the distortion, however, depends on how drastic the change in the first point is compared to the rest of the data. Thus zeroing the last point of a FID has little effect, because this point was presumably very close to zero (and contains no signal) to begin with.

Data are generally collected on NMR spectrometers in one of two modes: single phase or quadrature phase. Single-phase data produced by a single-channel spectrometer consist entirely of real components. The data acquired by quadrature detection, on the other hand, are complex. These detection schemes require slightly different FFT algorithms. The FFT algorithm is designed to transform $2n$ ($n = 1, 2, 3, ...$) complex data points. Consequently, quadrature FIDs are easy to transform. Fast FTs can be performed on single-phase data sets if the data set is first converted to a complex form. New data are not created, because N single-phase data points become $N/2$ complex data points. A typical mechanism

(Brigham, 1974) for Fourier-transforming single-phase data is first to shuffle the data so that each sequential pair of real data points becomes a complex data point (pair). Whether even-numbered points become real components or imaginary components is not important. A complex FFT is then performed. After the FFT, the data must be corrected using the algorithm given in Eqs. (5) and (6).

$$S_{Re}(n) = Re_n + Re_{M-n} + \cos \frac{\pi n}{N} (Im_n + Im_{M-n})$$

$$- \sin \frac{\pi n}{N} (Re_n - Re_{M-n}) \tag{5}$$

$$S_{Im}(n) = Im_n + Im_{M-n} - \sin \frac{\pi n}{N} (Im_n + Im_{M-n})$$

$$- \cos \frac{\pi n}{N} (Re_n - Re_{M-n}) \tag{6}$$

for $n = 0, 1, ..., N - 1$, where $M = N/2$ and N is the number of complex data points.

As a result of the multiplicative nature of the FFT, algorithms for fixed-point (integer) FFTs must constantly check for potential computer word overflow during transformation. When a potential overflow situation is detected, the data are scaled back, usually by dividing all data by 2. Because the amount of scaling is data-dependent and varies from spectrum to spectrum, scaling must be taken into account when spectra are compared with each other. Floating point FFTs do not need to check for overflow because of the large dynamic range of floating point representations. In fact, floating point numbers are self-scaling because the mantissa is restricted to values within one order of magnitude and the exponent is made to reflect the scale of the number. An external scaling parameter is not needed, and therefore spectra generated by floating point FFTs can be directly compared.

B. Post-FT Processing

After FT the spectroscopist has a complex spectrum whose absorption and dispersion mode signals are arbitrarily mixed in the real and imaginary data arrays. This results in a spectrum having peaks with seemingly arbitrary phases. Thus some peaks may appear to be inverted or have signal intensities both above and below the baseline. Nuclear magnetic resonance spectra are normally presented in the absorption mode, and a variety of methods exists for separating absorption components from dispersive ones.

1. Phasing

Whenever possible, phasing is the method of choice for obtaining absorption mode peaks. Phasing is superior to other techniques because it does not change the data content; it only transforms the data to a different format. Phasing is also reversible, so that the original data can be regenerated if necessary. In addition, the phasing algorithm does not modify the line shape of the absorption mode signal. Thus the FT of an exponentially decaying sine wave is a true Lorentzian line after phasing.

Phase shifts are introduced into the data by the spectrometer. These phase shifts are either independent of frequency or linearly dependent on frequency. Contributions to the constant, or so-called zero-order, phase distortion are generated by all the electronic components between the sample and the digitizer. Even coaxial cables introduce a zero-order phase shift because they have a finite length and a certain speed of propagation (typically 60% of the speed of light). Since the zero-order phase distortion is independent of frequency, all signals are equally affected. Furthermore, a zero-order phase shift of X degrees is indistinguishable from a zero-order phase shift of $X \pm n(360)$ degrees.

First-order or linear phase distortions are also introduced by the instrument. Filters (typically Butterworth) placed before the digitizer generally have a linear phase dependence on frequency, but the major source of first-order phase shifts is the delay between the end of the acquisition pulse and the start of data acquisition. Immediately after a pulse all the signals in the spectrum have the same phase. The delay between the pulse and the start of acquisition affects these signals differently. For example, the delay has little effect on the digitization of a low-frequency signal. A higher frequency signal, however, may go through more than a full cycle before digitization starts. A first-order phase shift of X degrees is not equal to an $(X + 360)$-degree phase shift.

The phasing algorithm is given in Eqs. (7) and (8).

$$S_{Re}(i) = Re(i)\cos(A + B_i) - Im(i)\sin(A + B_i) \tag{7}$$

$$S_{Im}(i) = Re(i)\sin(A + B_i) + Im(i)\cos(A + B_i) \tag{8}$$

where A is the zero-order correction and $B_i = (i - 1)/N$ multiplied by the first-order phase correction.

2. Magnitude Calculation

A simple procedure for ensuring that all peaks become upright is to calculate the magnitude of each complex data point. This technique has

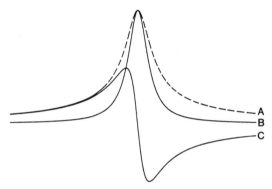

Fig. 2 Curve A shows the magnitude spectrum calculated from B and C; curve B the absorption mode of a Lorentzian line; and curve C the dispersion mode of the same line.

the advantage of being able to generate pseudo-absorption-mode spectra regardless of the phase of the original spectrum (Fig. 2). Magnitude calculation has its drawbacks, however, and it is usually used only when phasing is impractical (some two-dimensional experiments, for example). Magnitude calculations are not reversible, and, once performed, the imaginary data are no longer useful. In addition, the absorption and dispersion components are mixed in the calculation so that Lorentzian line shapes are severely distorted at the baseline. Gaussian line shapes, on the other hand, are not distorted by magnitude calculation. The algorithm for generating a magnitude spectrum $S_{mag}(i)$ is given in Eq. (9).

$$S_{mag}(i) = [Re^2(i) + Im^2(i)]^{1/2} \quad \text{for} \quad i = 1, 2, ..., N \quad (9)$$

A power spectrum can be calculated by squaring the magnitude spectrum; the power spectrum, however, is rarely used.

3. The Pseudoecho Technique

The pseudoecho method was originally developed (Bax *et al.*, 1981) to circumvent the problems associated with magnitude calculation for two-dimensional NMR data sets. Although quite useful, it suffers from severe limitations. The first step in the pseudoecho procedure is to apply an increasing exponential weighting function to the FID. If the proper assumption of line width is chosen, the natural decay of the FID will be exactly canceled. Next, a Gaussian weighting function centered around the midpoint of the FID is applied. Both functions can be combined and described in terms of time to give

$$g(t) = \exp(t/T_2) \exp[-(t - T_c)^2(3\pi b)^2/25] \quad (10)$$

where b is the desired Gaussian line width, T_2 the exponential decay time, and T_c the time at the center of the FID. The shape of the FID is now reminiscent of a spin echo, hence the name "pseudoecho." The data are symmetric about the center point. Consequently, FT of the data yields symmetric peaks that have only absorption components. If the magnitude spectrum is then calculated, pure Gaussian absorption mode peaks will be created. The primary disadvantage of this method is a severe decrease in the signal/noise ratio due to deemphasis of the first part of the FID (where most of the signal can be found). Also, the assumption of a single line width for the entire spectrum is troublesome, because spectra often consist of peaks with differing line widths.

4. Plotting

After data acquisition, apodization, FT, and phasing, the only fundamental operation left is plotting. Some NMR spectrometers still use analog plotters driven by digital-to-analog converters embedded in the spectrometer computers, but digital plotters have become available for most commercial NMR spectrometers, adding new features such as full plot annotation and automatic paper feed for multiple plots. Newer microprocessor-based digital plotters offer an additional advantage by performing many primitive operations that would otherwise occupy the spectrometer computer (e.g., drawing vectors and alphanumeric character generation).

III. Extended Features of FT NMR Processing

In addition to the basic computing functions outlined in Section I, several other features prove to be valuable. The scope of these extended features is limited only by the power of the computer and the ingenuity of the programmer.

A. Baseline Identification

To locate and quantify resonance lines in a spectrum, the nature of the baseline must be estimated. Some software systems implicitly estimate the offset and noise level of the baseline by assuming the offset is zero and the noise is some fraction of the largest peak. Other systems actually estimate the noise level by an iterative algorithm such as the one given in the following discussion.

The smallest value for a statistically significant signal against a noise background is generally given as three times the standard deviation of the noise (de Galan, 1971). Suppose the standard deviation and average of all the data (both signal and noise) are calculated. A first approximation of the noise level is then three times the standard deviation. Likewise, the offset is approximated by the average. The standard deviation can then be recalculated for all points suspected of being noise. A data point is defined as noise when

$$3(\text{standard deviation}) > \text{absolute value of (data} - \text{offset)} \qquad (11)$$

The old and new standard deviations can be compared and, if their difference is not within a predetermined limit, the process can be repeated with the new values for the offset and noise level. An iteration count is usually kept to stop the calculation in case subsequent standard deviations do not converge.

B. Peak Identification

Armed with the noise level and offset, location of the peaks is possible. Peaks can be defined in a variety of ways but, regardless of the criterion used, it is prudent to distinguish noise from possible peaks. A threshold can be defined as the offset plus the noise level. Only data points whose magnitudes are greater than the noise level need be considered in the search for possible peaks. The actual peak-picking algorithm should be able to select unresolved peaks and ignore noise spikes on peak shoulders. Cooper (1977) has outlined a good peak identification technique. His method searches for a local maximum in the data above the threshold. A peak is defined as a significant rise in the data followed by a significant fall. Only a point whose difference from the previous point is greater than the noise level is considered significant. This technique works well for most types of peaks as long as the threshold is chosen properly.

The procedure outlined merely finds the data point of greatest intensity. Rarely is this point located at the center of the peak. Accurate measurements of peak position and peak intensity can be obtained by interpolating the points in the immediate vicinity of the maximum point with an appropriate function. Linear interpolations yield good approximations for peak position but usually overestimate peak intensity. Parabolic interpolations, on the other hand, provide good approximations for both peak intensity and location.

C. Peak Integrations

If all the line widths and line shapes of a NMR spectrum are identical, then the height of each peak is proportional to the concentration of the chemical species giving rise to each signal. In almost all cases, line widths and line shapes are subject to a variety of chemical and instrumentation influences, however, and peak integrals must be measured for quantitative results.

Nuclear magnetic resonance lines theoretically have a Lorentzian distribution (Fig. 3).

$$f(\nu) = A\alpha/[\alpha^2 + (2\nu)^2] \tag{12}$$

where α is the full width at half-maximum height and A the peak intensity times α. Integrating between the limits $-\nu_1$ and $+\nu_1$ gives

$$\int_{-\nu_1}^{+\nu_1} f(\nu) \, d\nu = 2A\alpha \int_0^{\nu_1} d\nu/[\alpha^2 + (2\nu)^2] \tag{13}$$

$$= A \tan^{-1}(2\nu_1)/\alpha \tag{14}$$

When ν_1 is infinite, the total integrated intensity becomes

$$\int_{-\infty}^{+\infty} f(\nu) \, d\nu = A\pi/2 \tag{15}$$

It is unreasonable to integrate digitally over all frequencies, so an approximation must be made by choosing integration limits. Table I shows how the limits can be chosen as a function of peak width to obtain various accuracies for a Lorentzian line.

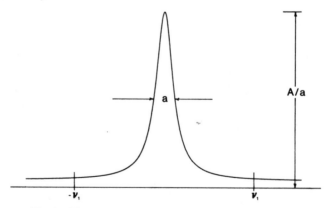

Fig. 3 The Lorentzian line shape, $f(\nu) = Aa/[a^2 + (2\nu)^2]$.

TABLE I

Accuracy of Integration for Lorentzian Lines
as a Function of Integration Range

Accuracy (% of Total Integral)	Integration Range (In Units of Line Width)
99.99	6366
99.90	636
99.00	63.6
90.00	6.31
80.00	3.08
50.00	1.00

Because most of the integrated intensity lies within a few multiples of the line width, small errors in the determination of the line width or the line shape have little effect on the integral so long as the limits are consistently chosen in terms of the individual line widths.

Automatic integrations can be implemented in peak analysis software if line widths are determined. The user should be given the option to change the default limit criterion, however, to integrate partially resolved peaks. The number of integrated points can be increased without changing the limit criterion by zero filling the data set while in the Fourier codomain. This action increases the digital resolution, or number of points per peak, thus increasing the precision of the integration. Convolution techniques, such as line broadening, do not affect integrals when the integration limits are chosen as a function of line width. However, sloping, offset, or distorted baselines can exert disastrous effects on integrals. In commercial spectrometers these problems are obviated by operator-controlled drift controls (two to four knobs). This method works reasonably well but is subject to operator bias. For precise integration it is better to correct baseline abnormalities prior to the integration procedure, using a statistically based procedure. The signal/noise ratio of a peak affects the precision of an integration, but the effect is usually negligible if enough data points are integrated and the signal/noise ratio of the peak is greater than about 5 or 6.

Integrals can be automatically calculated using a number of standard techniques. A first approximation can be obtained by summing all the points in the integration range. More precise results can be obtained by linearly interpolating the data points (trapezoidal rule). Integration by Simpson's rule (described in most calculus texts) is the method of choice, however, because it is based on a second-order polynomial interpolation.

If the theoretical line shape is known for a peak, the function can be fitted to the data, and the intensity, width, and integral can be determined. In practice, however, experimental line shapes can be too complex to characterize conveniently. A number of experimental factors can contribute non-Lorentzian components to the observed line shape.

D. Line Width Measurements

Line widths can be calculated before and after FT. The weighted average line width of a FID can be calculated by finding T_2^* and applying Eq. (17). T_2^* can be measured by performing a linear least squares operation on the natural logarithm of the magnitude of each complex point in the FID. The slope of the calculated line is simply the negative reciprocal of T_2^*.

$$\log_e[\text{magnitude}(i)] = -(i - 1)/T_2^* \qquad \text{for} \quad i = 1, 2, 3, ..., N \quad (16)$$

and

$$\text{line width} = 1/\pi T_2^* \qquad (17)$$

Line widths of individual peaks can be measured after FT using a straightforward algorithm. Once a peak has been located, its intensity can be best measured by a parabolic interpolation of the three most intense data points (many extended routines can use an estimated height calculated in this manner). Once the peak height is known, the data point on one side of the peak maximum closest to the half-height is found. A parabola is then fit to this point and its two nearest neighbors. The interpolated location at half-height is determined by finding the abscissa of the point on the parabola whose ordinate is the half-height. The procedure is repeated for the points closest to the half-height on the other side of the peak, and the line width is simply the difference between the interpolated locations. This method provides a reasonable approximation of line width only if the peak is sufficiently resolved from nearby peaks. Line widths for unresolved peaks can be determined by deconvolution or line-fitting methods.

IV. Special Problems of Broadline Spectra

Nuclei with spin greater than $\frac{1}{2}$ frequently yield broadened spectra that can present special problems. These problems include instrumentation limitations (sweep width, pulse excitation spectrum, etc.) and phenomenological limitations (poor resolution, quadrupolar effects, etc.). Nevertheless, many techniques developed originally for spin-$\frac{1}{2}$ spectra can be applied to wideline spectra; the methods available to spectroscopists interested in quadrupolar nuclei NMR are steadily increasing in number.

A. Baseline Identification and Correction

Baseline distortions can be caused by a variety of instrumentation and processing conditions. Ideal spectral baselines have an average mathematical value of zero over all segments of data. Distorted baselines, on the other hand, have varying local averages. Truncation of discrete data, for example, can cause serious distortions in the FT codomain. Experimental conditions such as overloaded amplifiers or digitizers can also distort baselines. A common cause of distortion in broadline spectra is pulse ringing during the acquisition of the first few data points of a FID. Often a distortion is not really a distortion at all, but an additional signal due to the sample matrix or an interfering substance. If the distortion cannot be prevented, it is useful to be able to compensate for it during processing. Well-characterized distortions due to spectral interference or matrix background can often be measured in the absence of the signal of interest. This spectral blank can then be subtracted from the distorted spectrum. However, baseline distortions due to processing factors and instrumentation problems can be highly variable, changing from spectrum to spectrum.

Pearson (1977) has written a routine for removing broad spectral features such as baseline distortions in the presence of narrow features such as signals. His algorithm fits a function to the signal-free components of a spectrum. The function is then subtracted from the data set to give a partially corrected spectrum. The process is repeated until the baseline is flat. Almost any function can be used for the fitting procedure. Since most distortions manifest themselves as continuous curves, fitting a multiorder polynomial is appropriate. The choice of data points for the fitting procedure is similar to the technique used in the automatic noise determination algorithm described earlier: Only data points whose values are less than three times the standard deviation of the entire data set are considered in the fit.

Pearson's method can be made more versatile if the user is allowed to specify the order of the correction. A zero-order polynomial correction, for example, is more useful than a sixth-order correction if the user wishes to change the spectral offset without risking modification of line shapes. Furthermore, the extra computation required for higher order corrections is unnecessary when the data have simple distortions. Very complex distortions can be corrected if an additional improvement is made in the algorithm so that the fitting procedure independently acts on small portions or blocks of the data set. Thus a fourth-order correction on four individual blocks of data can be as effective as a sixteenth-order correction on the entire spectrum. It is important that the user place the end points of each block in a signal-free region. Otherwise, part of a signal may contribute to the fitted polynomial and a discontinuity in the connection may occur. Default settings for the polynomial order and number of blocks are useful, but the user should be allowed to optimize the execution conditions of the correction, because severe distortions can make the routine slower than a FT. An example of the application of this technique to severely distorted data is given in Fig. 4.

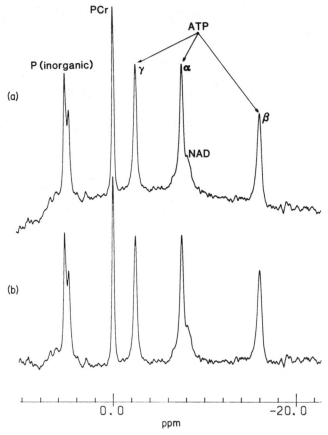

Fig. 4 ^{31}P-NMR spectrum of a beating perfused rat heart (Langendorff mode) before baseline flattening (a) and after a third-order correction over two blocks (b).

It should be pointed out that this baseline correction scheme cannot distinguish between a true signal and a distorted baseline. The distinction is simply between narrow and broad features of the spectrum. In practice a line width ratio (for baseline artifact to resonance line) exceeding 4 is necessary to avoid partial suppression of resonance lines, even when optimum conditions are set (correction order and block size). Therefore, for this algorithm to be effective, a broadline NMR spectrum must be presented in a narrowline format (i.e., the spectral peaks must be narrow with respect to the sweep width). This implies that extremely fast digitization rates may be needed.

B. Pulse Spectral Density Compensation

Because many broadline spectra also cover a large chemical shift range, large spectral windows must be excited and measured. To excite a large spectral window, very short pulses are needed. For these pulses to be effective, however, high-power amplifiers must be used. When the pulse excitation spectrum is sufficient to irradiate an entire broadline spectrum uniformly (Fig. 5A), the excitation is analogous to narrowline spectrum excitation. If the pulse is too long, however, the excitation spectrum will be nonuniform across the NMR spectrum, and in extreme cases it can even change phase (Fig. 5B). Fortunately, most instruments are capable of producing very uniform rectangular pulse shapes, and if the pulse shape is known, the excitation spectrum can be predicted and compensated for.

The excitation spectrum can be derived by Fourier analysis of the pulse. The pulse can be approximated as a sine wave multiplied by a delta function. Other factors such as noise and dc offset can be added, but their relative magnitude is so small that they can usually be ignored when calculating the excitation spectrum. The FT of the pulse then becomes a sinc function located at the frequency of the sine wave. The width of this function is inversely related to the length of the delta function or pulse width. Note that the sinc function becomes zero and changes sign at a frequency $N/$(pulse width), where N is any nonzero integer. The intensity of the sinc function (with respect to frequency) generated by a pulse of length T is

$$H(f) = AT[\sin(\pi Tf)/\pi Tf] \tag{18}$$

Given this relationship, it is possible to correct the data after FT by dividing the data by Eq. (18). Because the noise in the uncorrected spectrum is uniform over all frequencies, this correction increases noise levels in regions that correspond to the zero crossing points of the excitation

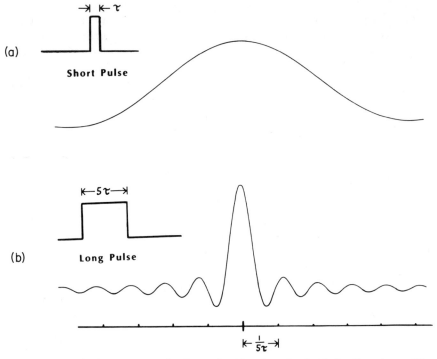

Fig. 5 The excitation spectrum of a short pulse (a) and a pulse that is five times longer (b).

spectrum. This side effect limits the usefulness of this correction method to cases where the pulse excitation spectrum is wide enough to cover the entire NMR spectrum without containing a zero crossing.

C. Convolution Difference

A widely used, simple method for separating wide spectral features from narrow ones is the convolution difference method. The basic ideas behind this algorithm are (a) that narrow features and broad features can be weighted independently before FT, and (b) that the difference between weighted and unweighted data can be calculated to extract narrow components selectively. For example, the T_2 of the broad components is estimated, and an exponential weighting function corresponding to this T_2 is applied to the FID. The weighted data contain only the broad signal components. Thus, when the weighted data are subtracted from the original data (either before or after FT), only the narrow components remain.

Likewise, the narrow-component spectrum can be subtracted from the original data to yield just the broad spectral features.

Although seemingly straightforward, this method is extremely dependent on the proper choice of the scaling factor applied to the weighted data during subtraction. Frequently, the spectroscopist must iterate several times to achieve a total separation of broad and narrow spectral features.

V. Current Trends in NMR Computations

In recent years we have seen the NMR spectrometer grow from a relatively simple, noncomputerized instrument to an extremely sophisticated instrument intimately involved with one or more computers. With advances in technology and the advent of even more complicated experiments, the NMR spectrometer is sure to evolve into an instrument which by today's standards will be as exotic as a 1980s NMR spectrometer would seem to a 1950s chemist. What form will the NMR spectrometer take 10 or 20 years from now? We can assume that advances in magnet technology will steadily increase the magnetic field strength and hence operational frequencies of the research NMR spectrometer (although interesting problems will be encountered). Far greater computerization of spectrometers and experiments is likely. It is conceivable (at least to the authors) that the future NMR spectrometer may function as a box into which the chemist inserts samples at one end and then collects interpreted structures at the other end. Shorter range developments are easier to predict, however, and a few likely features of future spectrometers are outlined here.

A. Multiprocessor Spectrometer Configurations

Already several manufacturers offer spectrometers with two built-in computers: one for data acquisition and spectrometer control and the other for data processing. Intelligent computer peripherals, such as plotters and high-speed graphics displays, further increase the number of processors in a NMR spectrometer. Other spectrometer designs may include additional computers dedicated to the tasks of sample changing, shimming, and spectrometer operations or diagnostics.

An additional concept available from several manufacturers is the NMR spectrometer work station. This work station is a second computer able to process data asynchronously and independently of the spectrometer console. Future work stations will probably become compatible with a wide range of peripherals, including high-density disks (perhaps shared), new output devices, and high-speed local area networks to handle the greatly increased information content of some of the newer NMR experiments (e.g., two-dimensional NMR).

An interesting concept for the spectroscopic laboratory of the future is the creation of a high-speed network of work stations. For example, one station could be optimized for two-dimensional NMR data processing with extensive amounts of fast memory and an integrated array processor. Other stations could be more modest in construction and be used for everyday one-dimensional NMR or other spectroscopic data processing. A third type of station could be dedicated to the control of a variety of hard copy peripherals, and a fourth dedicated to mass storage and network control. Work stations organized in this fashion should give increased laboratory throughput if properly designed and utilized.

B. Incorporation of New Technologies

We are currently seeing design changes due either directly or indirectly to the continuing drop in computer memory prices. These changes include the use of new graphic display technologies, larger data memories, and array processors.

1. High-Speed, High-Resolution Color Raster Scan Displays

Until recently, most spectrometers provided the user with a graphic output through digital-to-analog converters and an xy scope. This arrangement has the advantage of high resolution but suffers from an inability to maintain a bright, flicker-free display and the inability to provide color output. Raster scan display systems, on the other hand, provide very bright, steady graphics regardless of how much information is displayed. This is accomplished by having each display point on the video screen controlled by a unique bit (or set of bits for color) in the display processor's memory. The display processor is generally capable of rapidly executing simple functions such as vector and alphanumeric generation. Until recently the high cost of memory and the low resolution (up to 256×256) made raster scan display systems inappropriate for FT NMR spectrometer designs. Medium-resolution (512×512) and high-resolution display systems ($1K \times 1K$) are now widely available, but the price of high-resolution monitors is still high. Fortunately, medium-resolution color displays are very useful, because color can be effectively used to enhance the information shown on these display systems. Medium- and high-resolution color raster scan display peripherals are likely to become the display technology for future NMR spectrometers.

2. Larger Memories

A direct consequence of falling memory costs is the use of larger amounts of semiconductor memory in computers. In 1975 a typical NMR computer had an 8K data table; currently, in the early 1980s, the in-core data table is typically 32K or 64K words. Many spectrometers already accommodate 128K or larger data sets. Current NMR computers are designed with word sizes greater than 16 bits to minimize dynamic range constraints, and thus they can address theoretically large amounts of memory (one or more megawords). Large physical memories permit more efficient processing of large data sets (particularly two-dimensional NMR data sets). Large memories also allow real-time manipulation of multiple spectra (real-time spectral addition and subtraction, for example) by avoiding the inefficiencies of input and output. Multiuser systems in particular benefit when all users can share a physical memory without constant swapping to disk.

3. Raster-Type Plotters

The original analog plotters for NMR spectrometers were sufficient for the simple presentation of one-dimensional NMR data. Higher speeds and plot annotation have been achieved using digital pen plotters. Even fast digital pen plotters are not really adequate for the display of two-dimensional data sets. Raster plotters, on the other hand, appear to be very promising.

Several technologies exist for the hard copy production of raster-type graphics. The oldest and best established technology is the electrostatic plotter. Electrostatic plotters require special paper and a substantial amount of maintenance. Perhaps the most cost-effective raster-type plotter is the impact printer, but such printers are generally noisy and have a lower resolution than other plotters. Colored ink jet plotters are practical, and laser printers have been developed to provide high-resolution plots on ordinary paper. So far, laser printers are very expensive and are used only in mainframe computers or business environments. Their capability and speed, however, are impressive, and if laser printers become substantially less expensive, they may become the hard copy mechanism of future spectrometer systems.

C. Array Processors

Certain computing functions can be performed in a pipelined manner. Computations on arrays, for example, frequently can be performed in parallel. In other words, each data point in the array can be independently and simultaneously manipulated. In the logical extreme, one processor can be allotted to each data point in the array. Obviously this latter approach, while possible, is not very practical. An alternative design is to have several processors arranged so that the result of one processor is passed on to the next. While the second processor is performing the second step of a calculation, the first processor can be performing the first step of a computation on the next data point. This method can be extended to any number of processors and used for any number of data points, depending on the hardware.

This type of hardware is frequently called an array processor or vector processor. Array processors are typically difficult to program and difficult to interface efficiently to a host computer. The advantages obtained by an array processor are substantial, however, and NMR spectrometers are sure to use them.

Array processors can greatly accelerate array operations of a computer system. Since most NMR processing is array-oriented, significant increases in overall speed can be obtained. The FFT algorithm, for example, can be greatly accelerated by an array processor. Other operations such as apodizations and phasing are also well suited to implementation on an array processor. The increase in speed is great enough that Fourier transformation of each FID is possible during an experiment. This means that data can be collected in the frequency domain. Thus data overflow due to a large signal (perhaps solvent resonance) is unimportant. Another exciting possibility is the real-time FFT of the lock signal, so that shimming can be done by optimizing the line shape of the lock resonance. In addition to new methods of experiment and spectrometer control, array processors will accelerate ordinary processing and greatly improve the throughput of the NMR spectrometer.

D. Software Design

The nature of software for NMR spectrometers has evolved alongside hardware. Ten years ago, commercial software was written in assembly language, and modification by the spectroscopist was impossible. Modern spectroscopic software, on the other hand, is usually written in a higher level language. In addition, most commercial systems now allow the user to create and add new software modules to existing systems. These improvements are a result of more powerful computer hardware and advances in software engineering.

Only recently have the definitions and practices of good software programming been formulated. Today's programmer, unlike programmers in the past, takes a systematic, structured approach to software design. Desirable features of modern software include a high degree of modularity and structure. Modular systems are easier to create and modify as long as each module is independent. A consistent structure or style for all modules benefits both the user and the programmer by providing a uniform operating environment.

FORTRAN will probably continue to be the computer language used by most scientists, but newer languages are developing their own followers. For example, Pascal is popular (Cooper, 1981) because it guides the programmer into using good programming practices. Fortunately, good programming practices are easily adopted in FORTRAN, but the programmer must be aware of acceptable practices. Since the majority of spectroscopists are not professional programmers, poor software design has been a problem. Newer languages and a growing awareness of what constitutes good software should result in higher quality software in the future.

Acknowledgment

The authors acknowledge financial support from the Division of Research Resources, NIH (grant RR-01317).

References

Bartholdi, E., and Ernst, R. R. (1973). *J. Magn. Reson.* **11**, 9–19.

Bax, A., Freeman, R., and Morris, G. (1981). *J. Magn. Reson.* **43**, 333–338.

Brigham, E. O. (1974). "The Fast Fourier Transform," p. 169. Prentice-Hall, Englewood Cliffs, New Jersey.

Cooley, J. W., and Tukey, J. W. (1965). *Math. Computation* **19**, 297–301.

Cooper, J. W. (1977). "The Minicomputer in the Laboratory." Wiley (Interscience), New York.

Cooper, J. W. (1981), "Introduction to Pascal for Scientists." Wiley (Interscience), New York.

de Galan, L. (1971). "Analytical Spectrometry." Higler, London.

DeMarco, A., and Wuthrich K. (1976). *J. Magn. Reson.* **24**, 201–204.

Ferrige, A., and Lindon, J. (1978). *J. Magn. Reson.* **31**, 337.

Harris, R., and Mann, B. (1978). "NMR and the Periodic Table." Academic Press, New York.

Hoult, D. I., and Richards, R. E. (1974). *Proc. R. Soc. London* **344,** 311–340.

Hoult, D. I., and Richards, R. E. (1976). *J. Magn. Reson.* **15,** 484–497.

Levy, G., Terpstra, D., and Dumoulin C. (1981). *In* "Computer Networks in the Chemical Laboratory" (G. C. Levy, and D. K. Terpstra, eds.). Wiley (Interscience), New York.

Lindon, J., and Ferrige, A. (1980). *Prog. NMR Spectrosc.* **14,** 27–66.

Morris, G., and Freeman, R. (1979). *J. Am. Chem. Soc.* **101,** 760–762.

Pearson, G. A. (1977). *J. Magn. Reson.* **27,** 265–272.

Pines, A., Gibby, M., and Waugh, J. (1973). *J. Chem. Phys.* **59,** 569–590.

4 Factors Contributing to the Observed Chemical Shifts of Heavy Nuclei

G. A. Webb

Department of Chemistry
University of Surrey
Guildford, Surrey, United Kingdom

I. Introduction

In principle the factors that determine the measured chemical shifts of heavy nuclei are the same as those for light nuclei. In practice the relative importance of some of these factors changes on passing from lighter to heavier nuclei.

A chemical shift is a shielding difference between two nuclei of the same species in different environments. The nuclei concerned may belong to the same or to different molecules. Nuclear shielding, from the applied magnetic NMR field, is due to the electrons and nuclei present in the vicinity of the nucleus in question. Hence a basic description of the various factors responsible for chemical shifts requires a knowledge of the electronic environment of the nucleus concerned.

In spite of the many approximations involved, it is generally accepted that the most satisfactory current description of molecular electronic structure is that provided by molecular orbital (MO) theory (Pople and

NMR OF NEWLY ACCESSIBLE NUCLEI, VOL. 1

Beveridge, 1970). The more sophisticated ab initio MO calculations are usually performed on small molecules containing light atoms. In general, less precise, semiempirical MO descriptions are available for larger molecules and for those containing nuclei up to and including the first transition series of elements.

Although they have the advantage of being less demanding on computer time, the results from semiempirical calculations tend to be less satisfactory than those from ab initio calculations.

Molecular orbital calculations involving heavy atoms tend to be rather rare. Not only are they expensive to perform, but they should encompass interactions such as spin–orbit coupling that are usually neglected in the MO description of lighter atoms. Relativistic MO calculations include the effect of spin–orbit coupling (Pyykkö, 1978). However, relativistic MO calculations of nuclear shielding do not appear to have been reported to date.

It is perhaps of significance that the relative anisotropies of the reduced one bond couplings of Cd—C and Hg—C in $Cd(CH_3)_2$ and $Hg(CH_3)_2$ reveal relativistic contributions of 7.0 and 21.9%, respectively (Jokisaari et al., 1978). The magnitudes of these contributions suggest that any future serious MO description of the shielding of heavy nuclei should include relativistic effects, because these depend critically upon electronic wave functions close to the nuclei.

At the present time the majority of MO calculations of the factors influencing nuclear shielding are performed on molecules containing only light atoms (Ebraheem and Webb, 1978). Such calculations provide reasonable accounts of the various factors contributing to nuclear shielding. Before turning to a more detailed account of the molecular models available for nuclear shielding descriptions let us consider the approximate shielding ranges observed for some NMR nuclei.

It must be emphasized that the shielding range data presented in Table I reflect a number of features other than those of the intrinsic shielding ranges of the nuclei concerned. For example, some nuclei have been extensively studied for a number of years, whereas others have not. This may be due to factors such as relative NMR sensitivity of the nuclei concerned, stability and solubility of the molecules involved, and the variable amount of interest in particular areas of chemistry. In addition to NMR sensitivity and natural abundance, rapid relaxation, e.g., in quadrupolar nuclei or paramagnetic systems, may render the resonance signals more difficult to detect. Thus the results in Table I are probably deficient for well-understood reasons.

TABLE I

Some Reported Approximate Nuclear Shielding Ranges[a]

Nucleus	Shielding Range	Nucleus	Shielding Range
H	20	Sc	190
Li	10	Ti	1,700
B	200	V	7,600
C	650	Cr	1,800
N	1,000	Mn	3,000
O	1,500	Fe	1,300
F	800	Co	18,000
Na	20	Cu	800
Al	500	Zn	100
Si	400	As	650
P	700	Se	1,800
S	650	Nb	2,300
Cl	1,000	Mo	2,400
K	50	Rh	5,500
Ca	20	Ag	900
Ga	1,200	Cd	800
Ge	1,200	In	1,100
		Sn	2,300
		Sb	3,500
		Te	3,200
		Xe	7,400
		Cs	130
		La	300
		W	6,000
		Pt	13,000
		Hg	3,600
		Tl	7,000
		Pb	10,000

[a] Values are in parts per million.

All elements up to and including lead have stable NMR isotopes, with the exception of Tc, Pm, Ar, and Ce (Harris and Mann, 1978). Consequently most of the gaps in Table I will eventually be filled, and many of the shielding ranges presented will require updating as NMR applications proceed apace. With these shortcomings in mind, the data in Table I permit some general conclusions to be drawn.

First, the shielding ranges for the nuclei of groups I and II tend to be much smaller than those for members of the other groups of the periodic table. Second, the general trend is for heavier nuclei to have larger chemical shift ranges. There are some obvious exceptions to this conclusion. For example, cobalt, a member of the first transition series, appears to have the largest reported chemical shift range. On the other hand it is interesting to note that xenon has a much larger range than nuclei such as carbon and nitrogen. This suggests that the shielding ranges do not simply reflect the extent of the chemistry of the elements concerned.

Such general observations are broadly consistent with the fact that the elements of groups I and II have only s valence electrons in their ground states, are univalent as a rule, and tend to form mainly ionic compounds. In contrast, the other elements have valence electrons in orbitals with nonzero angular momentum such as p, d, or f and tend to engage in polyvalency as well as covalent bonding.

Usually polyvalency, and a greater polarizability of the valence electrons, becomes more common among the heavier elements. Often the compounds of heavier elements are more labile, thus chemical shift ranges can be very sensitive to changes in concentration, temperature, pH, and solvent. For example, the solvent-dependent shift range of Tl^+ salts exceeds 2400 ppm, whereas the anion and concentration dependences are in the range 10–100 ppm. Similarly, the concentration, anion, and pH dependence of Tl^{3+} covers a range of almost 2000 ppm (Hinton, 1982).

Thus there are a number of chemically familiar factors at work in the determination of nuclear shielding. It is not our intention to concentrate on the chemistry of the elements whose chemical shifts are of interest. Rather our aim is to provide an account of some of the models of the electronic interactions generally held responsible for the observed chemical shifts.

II. Intramolecular Factors

Some knowledge of the system under investigation will help to decide the relative importance of contributions to the observed chemical shifts from intra- and intermolecular interactions. Thus by assuming that the intermolecular effects are separated out we can turn our attention to intramolecular factors.

In calculations of nuclear shielding it is customary to consider a series of related molecules. Consequently, other intramolecular factors and possible intermolecular contributions may reasonably be regarded as being unchanged throughout the series.

A. Nuclear Magnetic Shielding Tensor

In an NMR experiment the field, B, experienced by a nucleus is related to the applied field B_0 by

$$B = B_0(1 - \sigma) \tag{1}$$

where σ, the shielding constant, represents the difference in environment for the nucleus under examination and for a bare nucleus. Hence σ is determined by the electronic distribution around the nucleus in question when other intra- and intermolecular effects are absent.

The shielding constant is one-third of the trace of a second-rank tensor σ. In principle NMR measurements on liquid crystals and solids can yield values for the individual components of the tensor σ and its anisotropy $\Delta\sigma$. In practice the results obtained from such measurements should be used with extreme caution because significant intermolecular contributions are often present.

However, if we assume that the measured results are for an isolated molecule then, for linear and symmetric top molecules,

$$\Delta\sigma = \sigma_{\parallel} - \sigma_{\perp} \tag{2}$$

where $\sigma_{\parallel}$ is the shielding component along the major molecular axis and $\sigma_{\perp}$ is that in the direction perpendicular to it. For less symmetric molecules $\Delta\sigma$ is usually defined by

$$\Delta\sigma = \sigma_{\alpha\alpha} - \tfrac{1}{2}(\sigma_{\beta\beta} + \sigma_{\gamma\gamma}) \tag{3}$$

where the σ_{ii}'s are the three principal tensor components taken such that $\sigma_{\alpha\alpha} \geqslant \sigma_{\beta\beta} \geqslant \sigma_{\gamma\gamma}$. From Eqs. (2) and (3) it follows that whether a given nucleus will have an isotropic or anisotropic shielding tensor depends on the nuclear site symmetry (Buckingham and Malm, 1971).

A nucleus at a tetrahedral or octahedral site has an isotropic shielding tensor, whereas lower symmetries produce anisotropy. For heavier atoms electronic distortions frequently occur, e.g., from the Jahn–Teller effect, thus nominally octahedral metal complexes frequently produce a lower site symmetry for the metal atom, e.g., D_{4h} or D_{3d}. Hence the anisotropy of the shielding tensor is a reflection of the symmetry of the actual nuclear environment. In addition the magnitude of $\Delta\sigma$ is often related to the size of the chemical shift range, such that heavier nuclei tend to have larger values of $\Delta\sigma$, e.g., square-planar Platinum(II) compounds have $\Delta\sigma$ values ranging from 2,500 to 10,500 ppm (Keller and Rupp, 1971).

B. Theoretical Models of Nuclear Shielding

Quantum mechanical calculations of the nuclear shielding tensor are usually couched in the formalism presented by Ramsey (Ramsey, 1950). This gives rise to paramagnetic σ^p and diamagnetic σ^d components of the shielding tensor:

$$\sigma = \sigma^d + \sigma^p \qquad (4)$$

where σ^d involves the concept of free rotation of electrons about the nucleus and σ^p refers to the hindrance to this rotation caused by other electrons and nuclei in the molecule. It should be kept clearly in mind that the terms "diamagnetic" and "paramagnetic" do not refer to the magnetic properties of the molecule under consideration but rather arise from the mathematical shapes of the expressions for σ^d and σ^p.

It is generally accepted that Ramsey's approach is of limited value in a discussion of molecular problems (Webb, 1978; Ando and Webb, 1983). There are a number of drawbacks to this approach, including the fact that σ^d and σ^p are of opposite sign, such that as the molecular size increases σ becomes the relatively small difference between two large, variable terms and thus is subject to considerable error.

A further difficulty is that a knowledge of all the excited states is required for the evaluation of σ^p. Finally, the values of σ^d and σ^p obtained depend upon the origin chosen for the calculation. Although this does not affect the value of σ, it leads to problems when comparing results from different sources.

In practice the choice of atomic orbitals used in the calculation is invariably an incomplete set. Under these conditions the value of σ also usually depends upon the choice of origin. This point is illustrated by a comparison of calculations for first- and second-row hydrides (Höller and Lischka, 1980). It is found to be necessary to include orbitals up to the $3d$ level to obtain results that are origin-independent for first-row hydrides; whereas $4d$ and $5d$ sets are required for second-row hydrides if origin independence of the results is to be achieved. The requirement of such large basis sets makes the treatment of larger molecules time-consuming and probably even prohibits their consideration. Since it is obvious from the experimental point of view that σ does not depend on the origin chosen, at the present time calculations employing Ramsey's approach are restricted to small molecules containing only light atoms.

A more applicable model of nuclear shielding of specific chemical interest is one in which a number of localized shielding contributions are considered. Such an approach allows nuclear shielding to be represented by a sum of expressions each of which can be related to particular atoms and bonds in a molecule (Saika and Slichter, 1954).

Pople (Pople 1962a, 1962b, 1964) developed a MO theory within the framework of an independent electron model to produce a shielding relation of the type

$$\sigma = \sigma^d(\text{loc}) + \sigma^d(\text{nonloc}) + \sigma^d(\text{inter})$$
$$+ \sigma^p(\text{loc}) + \sigma^p(\text{nonloc}) + \sigma^p(\text{inter}) \tag{5}$$

The various diamagnetic and paramagnetic terms in Eq. (5) are not directly comparable to their counterparts bearing these names in Ramsey's theory.

The local terms $\sigma^d(\text{loc})$ and $\sigma^p(\text{loc})$ arise from electronic currents localized on the atom containing the nucleus of interest; $\sigma^d(\text{nonloc})$ and $\sigma^p(\text{nonloc})$ are derived from currents on neighboring atoms, and $\sigma^d(\text{inter})$ and $\sigma^p(\text{inter})$ arise from shielding currents not localized on any atom in the molecule, e.g., ring currents. The two interatomic terms in Eq. (5) account for a shielding contribution of a few parts per million at most. As shown in Table I, most nuclei have shielding ranges of several hundred parts per million, so that shielding effects due to the anisotropy of the susceptibility tensor of neighboring atoms, as well as those from ring currents, are normally neglected when discussing chemical shift trends of heavy nuclei.

The difficulties related to origin dependence of the shielding results and to having σ represented by a small difference between two large terms, such as those noted with Ramsey's model, are absent when Pople's sum-over-states (SOS) approach is used together with gauge-invariant atomic orbitals (GIAOs) to evaluate the terms given in Eq. (5). The other undesirable aspect of Ramsey's formulation is the requirement for some knowledge of all the excited states, including those of the continuum, to evaluate the paramagnetic shielding contribution. In contrast, the paramagnetic terms in Eq. (5) are determined by molecular excited singlet states, information on which is more readily available. However, inclusion of the excited states is usually the least satisfactory aspect of SOS nuclear shielding calculations.

Ditchfield (1972, 1974) has presented a different formulation of the shielding tensor which again comprises a number of localized contributions. This model uses GIAOs and is derived from finite perturbation theory (FPT). It has the advantage of not requiring specific details on the excited molecular states. At the ab initio level FPT calculations provide very satisfactory shielding results for small molecules containing light atoms (Ditchfield 1972, 1974). Other workers have used the model developed by Ditchfield to obtain very promising shielding results for the nuclei in N-methylformamide (Ribas Prado et al., 1981a), imidazole (Ribas Prado et al., 1981b), and cytosine (Giessner-Prettre and Pullman, 1982).

An extension of this model to larger molecules has so far, necessitated the employment of semiempirical CNDO/INDO parameters (Garber et al., 1979; Ellis et al., 1980). By employing modified input MO parameters satisfactory ^{13}C and ^{11}B shielding results are obtained in spite of the truncated nature of the basis set used. From these results it is clear that multicenter terms play an important role in determining σ, particularly when multiple bonding occurs.

In contrast, Pople's SOS model with CNDO/INDO parameters consists of one-center expressions for the local and nonlocal shielding terms represented in Eq. (5). In spite of this limitation Pople's approach is the most widely adopted one for the calculation of nuclear shielding.

Within the confines of the approximations inherent to CNDO/INDO calculations, the rotationally averaged values of the local shielding terms for nucleus A, which is assumed to have only s and p valence electrons, are given by Eqs. (6) and (7).

$$\sigma_A^d(\text{loc}) = \frac{\mu_0 e^2}{12\pi m} \sum_B^A P_{\mu\mu}\langle\mu|r_{A\mu}^{-1}|\mu\rangle \tag{6}$$

$$\sigma_A^p(\text{loc}) = -\frac{\mu_0 e^2 \hbar^2}{6\pi m^2}\langle r^{-3}\rangle_{np}\sum_j^{\text{occ}}\sum_k^{\text{unocc}}(E_k - E_j)^{-1}$$

$$\times \left[(C_{yAj}C_{zAk} - C_{zAj}C_{yAk})\sum_B(C_{yBj}C_{zBk} - C_{zBj}C_{yBk})\right.$$

$$+ (C_{zAj}C_{xAk} - C_{xAj}C_{zAk})\sum_B(C_{zBj}C_{xBk} - C_{xBj}C_{zBk})$$

$$+ \left.(C_{xAj}C_{yAk} - C_{yAj}C_{xAk})\sum_B(C_{xBj}C_{yBk} - C_{yBj}C_{xBk})\right] \tag{7}$$

where $P_{\mu\mu}$ is the charge density in the atomic orbital μ that is at an average distance r_A from nucleus A, C_{xAj} is the LCAO coefficient of the P_x orbital on atom A in the MO j, etc. The summation over neighboring atoms B includes A. In Eq. (7), E_k and E_j refer to the energies of MOs k and j that are unoccupied and occupied, respectively, in the ground state. The excitation energies $E_k - E_j$ are for excited singlet states which are mixed by the applied magnetic field with the ground state. From the ordering of the LCAO coefficients in Eq. (7) it is apparent that $\sigma \rightarrow \sigma^*$, $\sigma \rightarrow \pi^*$, $n \rightarrow \sigma^*$, and $n \rightarrow \pi^*$ singlet transitions, but not $\pi \rightarrow \pi^*$ transitions, contribute to $\sigma_A^P(\text{loc})$. It is also clear that both atoms A and B must possess valence p electrons for $\sigma_A^P(\text{loc})$ to be nonzero, because $\langle r^{-3}\rangle_{np}$ represents the expectation value of the inverse cube of the separation of the p electrons with primary quantum number n from the nuclei in question.

Thus, within the SOS description, the nuclear shielding of the elements of groups I and II of the periodic table has no contribution from $\sigma^P(\text{loc})$. The small chemical shift ranges of these nuclei are thus understandable on the basis of small variations in the other terms occurring in Eq. (5). In general, such changes are considerably smaller than those observed in $\sigma^P(\text{loc})$ for nuclei that have nonzero values of $\sigma^P(\text{loc})$.

Expressions analogous to Eqs. (6) and (7) are available for the nonlocal shielding terms (Ando and Webb, 1983). These are obtained by assuming that the electrons on atom B produce a shielding effect at atom A, which may be approximated by replacing the induced moment of B by a point dipole (Pople, 1957). Thus the nonlocal shielding terms become one-center in nature.

Calculations of the nonlocal shielding expressions for first-row nuclei show that $\sigma^d(\text{nonloc})$ contributes less than 1 ppm to the total shielding, whereas the size of the contribution from $\sigma^P(\text{nonloc})$ depends on the multiplicity of the bonding involved (Webb, 1978). The largest value of $\sigma^P(\text{nonloc})$ found for first-row nuclei is about -12 ppm for the central nitrogen atom in the azide anion, N_3^-, and the corresponding value of $\sigma^P(\text{loc})$ is -261 ppm (Jallali-Heravi and Webb, 1978).

Consequently, within the confines of Pople's SOS formulation, nuclear shielding is largely described by $\sigma^d(\text{loc})$ and $\sigma^P(\text{loc})$, with $\sigma^P(\text{nonloc})$ being of marginal importance for multiply bonded light nuclei. Preliminary calculations for heavier nuclei suggest that both $\sigma^P(\text{loc})$ and $\sigma^P(\text{nonloc})$ are larger than for light nuclei but that the relative significance of $\sigma^P(\text{nonloc})$ remains small (Na-Lamphun and Webb, 1982). The small chemical shift range of hydrogen, and other atoms with only s valence electrons, is thus ascribed to the variation in $\sigma^d(\text{loc})$.

The matrix element in Eq. (6) is often evaluated from

$$\langle \mu | r_{A\mu}^{-1} | \mu \rangle = Z_\mu / n^2 a_0 \tag{8}$$

where Z is the effective nuclear charge for the atomic orbital μ and a_0 the Bohr radius. Thus $\sigma_A^d(\text{loc})$ is expected to vary periodically in much the same way as the effective nuclear charge. The value of $\sigma_A^d(\text{loc})$ is very similar to that of the free atom, the magnitude of which, as shown in Table II, increases dramatically with atomic mass (Malli and Froese, 1967). Calculations of Eqs. (6) and (8) using various semiempirical MO parameter sets show that, for given light and heavy nuclei in a range of molecular

TABLE II

Some Free-Atom Values of $\sigma_A{}^d$[a]

Nucleus	$\sigma_A{}^d$	Nucleus	$\sigma_A{}^d$
Li	101.45	Ni	2282.32
B	201.99	Cu	2400.71
C	260.74	As	2877.18
N	325.47	Se	2998.37
O	395.11	Br	3121.19
F	470.71	Mo	4000.64
Na	628.90	Ru	4265.53
Al	789.88	Pd	4536.25
Si	874.09	Ag	4673.85
P	961.14	Sn	5085.55
S	1050.47	I	5501.64
Cl	1142.64	Xe	5642.32
K	1329.36	Cs	5780.22
Ti	1622.72	Os	9064.44
V	1726.62	Ir	9229.54
Cr	1833.02	Pt	9395.58
Mn	1942.10	Au	9561.41
Fe	2052.91	Hg	9729.07
Co	2166.38	Tl	9894.16

[a] Values are in parts per million.

environments, $\sigma_A{}^d(loc)$ varies by less than 5% (Ebraheem et al., 1976; Na-Lamphun and Webb, 1982; Schmidt and Rehder, 1980). Consequently shielding variations of a given nucleus arise primarily from a change in the value of $\sigma_A{}^p(loc)$ in different molecular environments.

A change of less than 5% in the free-atom value of σ^d for hydrogen nuclei is too small to account for its restricted chemical shift range. Although this is not necessarily the case with the other group I and II nuclei, it points not only to the importance of ring currents and other special shielding influences for hydrogen nuclei but also to shortcomings in the SOS model when it is used in semiempirical calculations. Similar calculations, using the FPT model, reveal that two-center terms can be as important as one-center terms for proton chemical shifts and that three-center terms can also be of significance (Dobosh et al., 1979).

With a closer inspection of $\sigma_A{}^p(loc)$, Eq. (7) reveals three potential sources of variation, namely, $\langle r^{-3} \rangle_{np}$, the excitation energy $E_k - E_j$, and the MO coefficients of the p orbitals. Equation (9)

$$\langle r^{-3} \rangle_{np} = \tfrac{1}{3}(z_{np}/na_0)^3 \tag{9}$$

shows that $\langle r^{-3} \rangle_{np}$ depends critically on the atomic number. Some average

TABLE III

Some SCF Average
Values of $\langle r^{-3} \rangle_{np}$

Nucleus	$\langle r^{-3} \rangle_{np}$
B	0.775
C	1.692
N	3.101
O	4.974
F	7.546
Al	1.055
Si	2.041
P	3.319
S	4.814
Cl	6.709
Ga	2.867
Ge	4.785
As	6.987
Se	9.228
Br	11.876

[a] Values are in (atomic
units)$^{-3}$.

values of $\langle r^{-3} \rangle_{np}$, obtained from SCF calculations on neutral atoms, are presented in Table III (Whiffen, 1964). Naturally, for third-row and heavier nuclei, d orbitals should be included in the MO description, with f orbitals also being considered for the fourth row and beyond. An expression analogous to Eq. (7) has been derived for the d-orbital contribution to $\sigma_A^p(\text{loc})$ (Jameson and Gutowsky, 1964). Because the d-orbital shielding contribution also has a $\langle r^{-3} \rangle$ dependence, it shows a periodic variation similar to that of the p orbitals.

For nuclei of a given element, chemical shift changes due to environmental differences can arise from simultaneous variations in $\langle r^{-3} \rangle$ for the valence electrons, their excitation energies, and the appropriate MO coefficients. Often it is not feasible to separate these various changes. In cases where the chemical shift trend is dominated by one of these variables, a correlation is often possible between the chemical shift and another experimental parameter depending on the same variable, as discussed in Section II,G.

In closing this section mention should be made of the fact that Eq. (7) is often simplified by means of the closure approximation and the replacement of $E_k - E_j$ by ΔE. This is usually referred to as the average excitation energy (AEE) approximation, the appropriate value of the AEE being expressed by ΔE. Because ΔE corresponds to some largely unknown weighted average of singlet transition energies, it is usually not calculated a priori but treated as an empirical parameter.

Within the AEE approximation, Eq. (7) becomes

$$\sigma_A{}^P(\text{loc}) = -\frac{\mu_0 e^2 \hbar^2}{8\pi m^2} \frac{\langle r^{-3}\rangle_{np}}{\Delta E} \sum_B Q_{AB} \tag{10}$$

where

$$Q_{AB} = \tfrac{4}{3} \delta_{AB}(P_{xAxB} + P_{yAyB} + P_{zAzB})$$
$$- \tfrac{2}{3} (P_{xAxB}P_{yAyB} + P_{xAxB}P_{zAzB} + P_{yAyB}P_{zAzB})$$
$$+ \tfrac{2}{3} (P_{xAyB}P_{xByA} + P_{xAzB}P_{xBzA} + P_{yAzB}P_{yBzA}) \tag{11}$$

and δ_{AB} is the Kronecker delta and the P's are elements of the charge density bond order matrix. Thus the first set of terms on the right-hand side of Eq. (11) is a charge density expression, the remaining two sets being bond order expressions.

A modified form of Eq. (11), including d orbitals, is often employed for heavier nuclei. This involves restricting the summation of Q_{AB} to valence orbitals centered on atom A. The resulting expression for $\sigma_A{}^P(\text{loc})$ becomes

$$\sigma_A{}^P(\text{loc}) = (-\mu_0 e^2 \hbar^2 / 6\pi m^2 \, \Delta E)(\langle r^{-3}\rangle_{np} P_\mu + \langle r^{-3}\rangle_{nd} D_\mu) \tag{12}$$

where np and nd refer to the valence p and d electrons and P_μ and D_μ represent, respectively, the p and d electron imbalance about nucleus A. The expression for P_μ is

$$P_\mu = (P_{xx} + P_{yy} + P_{zz}) - \tfrac{1}{2}(P_{xx}P_{zz} + P_{yy}P_{xx} + P_{yy}P_{zz})$$
$$+ \tfrac{1}{2}(P_{xy}P_{yx} + P_{xz}P_{zx} + P_{zy}P_{yz}) \tag{13}$$

thus the maximum value of P_μ is 2, which corresponds to two p orbitals being filled and one empty. The maximum value of D_μ is 12, which occurs, for example, with the low-spin d^6 octahedral and d^4 tetrahedral electronic configurations.

C. Paramagnetic Effects

The presence of a paramagnetic center, either in the molecule of interest or in a neighboring molecule, can severely influence the observed nuclear shielding. There are two possible mechanisms that may be operative in producing the induced shielding effects, and the observed shielding may be influenced by either or both of the processes in question.

One of these gives rise to contact interactions as shown by Eq. (14), whereas the other produces the dipolar, or pseudocontact, interactions given by Eq. (15).

$$\Delta B/B_0 = - a_N g_e^2 \beta_e^2 S(S + 1)/3 g_N \beta_N KT \tag{14}$$

$$\Delta B/B_0 = - [(3 \cos^2 \theta - 1)\beta_e^2 S(S + 1)/3 r^3 KT]F(g) \tag{15}$$

In Eqs. (14) and (15) the paramagnetic induced resonance shift ΔB is expressed as a function of the applied magnetic field B_0. The hyperfine interaction constant a_N is expressed in tesla, g_e is the rotationally averaged electronic g value, β_e is the Bohr magneton, g_N and β_N are the corresponding nuclear parameters, and S denotes the spin of the unpaired electrons. In Eq. (15) r is the separation between the resonating nucleus and the unpaired electron, θ is the angle between the vector defined by r and the principal symmetry axis of the molecule, and $F(g)$ is an algebraic function of the g tensor components (Webb, 1975, Orrell, 1979). Substantial increases and decreases in nuclear shielding can be produced by either or both of the contact and dipolar interactions.

An example of an isoshielding contour map for a system with a d^1 electronic configuration in a strong octahedral crystal field is shown in Fig. 1. On this map both the contact and dipolar shielding contributions are indicated. It is clear from Fig. 1 that a complex pattern of induced shifts exists in various regions of a paramagnetic molecule and that shielding can change very rapidly as a function of distance (Golding and Stubbs, 1979).

A compact general expression has been reported for evaluation of the induced paramagnetic shielding effects for any d-electron system (Golding and Stubbs, 1980). Because a large number of compounds of elements with valence d and f electrons are paramagnetic, an understanding of paramagnetic influences on nuclear shielding is of importance when considering the NMR spectra of heavier nuclei.

D. Dynamic Effects

The effects of molecular vibration and rotation on nuclear shielding were first predicted by Ramsey (Ramsey, 1952). These effects may be observed in terms of a temperature dependence of the resonance position in the limit of zero pressure and by shielding changes due to isotopic substitution of neighboring nuclei as discussed in Section II,E. In effect, both these observations arise from a change in the average nuclear configuration of a molecule as it vibrates and rotates.

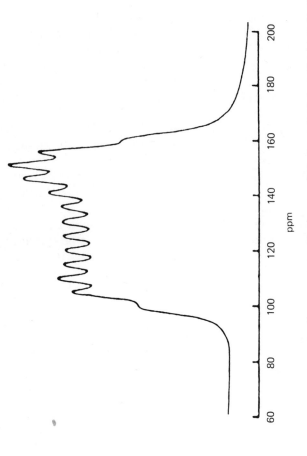

Fig. 1 ^{59}Co spectrum of approximately equal volumes of [Co(en)]Cl$_3$ in H$_2$O, 25% (v/v) D$_2$O–H$_2$O, 50% (v/v) D$_2$O–H$_2$O, 75% (v/v) D$_2$O–H$_2$O, and 99.8% D$_2$O mixed at 273 K and determined at 280 K (Bendall and Doddrell, 1978).

For a pure dilute gas the temperature dependence of the nuclear shielding may be described by a virial-type expansion:

$$\sigma(T, \rho) = \sigma_0(T) + \sigma_1(T)\rho + \sigma_2(T)\rho^2 + \cdots \qquad (16)$$

where $\sigma_0(T)$, the independent molecular shielding, is temperature-dependent because of averaging over intramolecular motions and the density-dependent terms ρ describe intermolecular effects which are dealt with in Section III.

Vibrational and rotational shielding contributions for gases tend to increase as the nuclear mass increases and can be appreciable, for example, 50 ppm for the ^{31}P shielding in PH_3 (Jameson and Jameson, 1978). Although the situation in solutions, where most NMR measurements are taken, is undoubtedly different, the point remains that calculated and experimental shielding values should be compared carefully. Until systematically discounted, dynamic effects on the shielding of heavier nuclei should not be ignored. For example, the shieldings of the ^{63}Cu, ^{35}Cl, ^{81}Br, and ^{127}I nuclei of cuprous halides are found to depend linearly on the temperature (Becker, 1978). This observation is accounted for in terms of an increase in the vibrational overlap with temperature. A further aspect of vibrational effects on magnetic shielding is provided by some diamagnetic cobalt(III) octahedral complexes (Juranic, 1981).

As the temperature increases, a larger occupancy of the excited vibrational levels occurs, and this causes a decrease in the effective ligand field splitting. As a consequence the ^{59}Co shielding has a linear dependence on the wavelength of the lowest $d \rightarrow d$ electronic transition.

E. Isotopic Effects of Neighboring Nuclei

Bond lengths and angles are dependent on isotopic mass because of anharmonic vibrations and centrifugal distortions. Isotopic substitution results in the heavier molecule lying lower on the potential curve, such that it experiences a smaller mean displacement from the equilibrium geometry.

In general, nuclear shielding decreases with bond extension, and thus the nuclei in the heavier molecule tend to be more shielded than those in the corresponding lighter molecule. The magnitude of the isotopic shift depends on the separation between the nucleus of interest and the site of substitution. Nuclei with large shielding ranges tend to have greater isotopic shifts. Consequently this effect assumes greater significance for heavier nuclei. For example, substitution of deuterium for hydrogen gives rise to a shielding change of 5.2 ppm per substituted atom for ^{59}Co (Bendall and Doddrell, 1978), whereas a corresponding change of less than 0.5 ppm is produced for ^{13}C (Grishin *et al.*, 1971).

Two other aspects of the influence of isotopic substitution on the observed magnitude of nuclear shielding are noteworthy. One is that the influence depends on the fractional change in mass on substitution; e.g., changing from ^{37}Cl to ^{35}Cl and from ^{81}Br to ^{79}Br produces induced ^{195}Pt shielding changes of 0.167 and 0.028 ppm per isotope, respectively (Ishmail *et al.*, 1980). Obviously substituting ^{3}H for ^{1}H produces the largest fractional change.

The final aspect is that the size of the isotopic shielding effect depends on the number of atoms replaced in a given molecule. An example of this is demonstrated for $[Co(en)_3]Cl$ in H_2O-D_2O (Bendall and Doddrell, 1978) in Fig. 1. The 13 observed ^{59}Co resonances correspond to the members of the isotopic homologous series in which the 12 exchangeable H atoms per molecule are replaced by D.

F. Heavy-Atom Effects

Another intramolecular influence of one nucleus on the shielding of another is the so-called heavy-atom effect. In this case the nuclear shielding of a lighter atom is influenced by a change in its electronic energy levels on account of spin-orbit coupling interactions derived from a neighboring heavy nucleus.

An example of this effect is provided by the ^{13}C chemical shifts of iodomethanes. Compared to methane the ^{13}C nuclei of CH_3I, CH_2I_2, CHI_3, and CI_4 are shielded by 18.5, 51.7, 137.6, and 290.2 ppm, respectively (Stothers, 1972).

Shielding calculations involving heavy-atom spin–orbit coupling by means of third-order perturbation theory account for the large shifts induced by the heavy atoms concerned. In the case of HI, spin–orbit coupling appears to contribute 51% of the total proton shielding (Morishima *et al.*, 1973).

Similar calculations on all the mono-halo-substituted methanes using optimized INDO parameters can reproduce the ^{13}C shieldings, including the nonadditivity of the shielding changes with increasing number of substituents (Cheremisin and Schastnev, 1980). The results reveal that the effect of spin–orbit coupling interactions is small for chlorine substitution, essential for bromine substitution, and predominant when iodine substitution occurs.

Because the magnitude of the spin–orbit coupling interaction depends approximately on the fourth power of the nuclear charge, it is apparent

Fig. 2 An isoshielding diagram for the case of NMR nucleus in the *xy* plane for a d^1 transition metal ion in a strong crystal field of octahedral symmetry, where bonding effects have been considered. Ligands are located ±0.2 nm from the central metal ion along the three axes. The contours of equal chemical shift $\Delta B/B$ are in parts per million. (A bar over the number indicates a negative value.)

that such effects are large for all heavy nuclei. A comprehensive relativistic theory of nuclear shielding would account for the effect of spin–orbit coupling on both the shielding of the heavy atom concerned and that of its lighter neighbors.

Other aspects of heavy atoms that await such a comprehensive theory are the inverse halogen dependence noted for the shielding of ^{47}Ti and certain other transition-metal nuclei (Kidd, 1980) and the apparent heavy-atom effect dependence on position in the periodic table. For example, ^{13}C shieldings do not appear to be influenced appreciably by the presence of group IVB elements (Laitem *et al.*, 1980), whereas group VB elements produce a decrease in ^{13}C shielding (Ashe, 1978) and group VIB elements an increase in ^{13}C shielding (Laitem *et al.*, 1980).

G. Correlations between Nuclear Shielding and Other Spectroscopic Observations

As demonstrated by Eqs. (2) and (3), the nuclear shielding tensor can be anisotropic. Although rapid molecular motion in solution normally removes the anisotropy from consideration when chemical shifts are under discussion, it can remain as an important contributor to nuclear relaxation. The shielding anisotropy contributions to the nuclear relaxation times T_1 and T_2 are given by T_{1SA} and T_{2SA}, respectively, where, in the motional narrowing limit,

$$1/T_{1SA} = \mu_0 \gamma_I^2 B_0^2 \, \Delta\sigma^2 \tau_c / 30\pi \tag{17}$$

and

$$1/T_{2SA} = 7\mu_0 \gamma_I^2 B_0^2 \, \Delta\sigma^2 \tau_c / 180\pi \tag{18}$$

where γ_I is the magnetogyric ratio of the relaxing nucleus and τ_c the molecular rotation correlation time. Equations (17) and (18) show the dependence of T_{1SA} and T_{2SA} on the applied magnetic field, which provides a way of detecting the presence of this relaxation process.

An example of the importance of shielding anisotropy as a relaxation mechanism is afforded by variable-temperature and -field studies on $Pb(CH_3)_3Cl$ (Hays *et al.*, 1981). When the applied field is 7.05 T, the ^{207}Pb T_1 relaxation is dominated by the shielding anisotropy mechanism, whereas the spin–rotation interaction controls it at 2.35 T. A further example is provided by $Hg(C_6H_5)_2$ (Gilles *et al.*, 1981); T_1 measurements of ^{199}Hg reveal an anisotropy of 6800 ± 680 ppm, which controls the relaxation.

Shielding anisotropy may prove to be a serious limiting factor in the promulgation of high-field NMR techniques for heavy nuclei. The gain in sensitivity, available at higher applied fields, may well be mitigated by line broadening due to rapid relaxation. In this regard, the study of nuclei whose line broadening is due to a non-field-dependent interaction, e.g., quadrupolar relaxation, stands to gain more from the use of high-field NMR techniques.

Another relationship between nuclear shielding and relaxation may be found for quadrupolar nuclei (Deverell, 1969). The nuclear quadrupole coupling constant χ depends on the electric field gradient at the nucleus eq

and the nuclear quadrupole moment eQ, as shown by Eq. (19)

$$\chi = eqeQ/h \tag{19}$$

For p valence electrons eq depends on $\langle r^{-3} \rangle_{np}$; thus from a comparison of Eqs. (7) and (19) it is apparent that

$$\sigma_A^p(\text{loc}) \propto \chi \tag{20}$$

If ΔE is considered constant, for a series of molecules, a change in χ is reflected by a shielding variation. This has been demonstrated for a number of compounds including some inorganic chlorides (Johnson, *et al.*, 1969).

When quadrupolar relaxation is the dominating process, which is usually the case for nuclei with $I > \frac{1}{2}$, the resonance signal half-width $\Delta_{1/2}$ for axially symmetric systems depends on χ according to

$$\Delta_{1/2} \propto \chi^2 \tau_c \tag{21}$$

Thus for a series of molecules containing a quadrupolar nucleus,

$$(\Delta_{1/2})^{1/2} \propto \sigma_A^p(\text{loc}) \tau_c^{1/2} \tag{22}$$

In the derivation of expression (22) it is assumed that the shielding differences for the quadrupolar nucleus in the various molecules considered arise solely from a change in $\langle r^{-3} \rangle_{np}$, that the solution is of unit viscosity, and that axial electric symmetry is maintained at the nucleus in question.

Under these circumstances $\sigma_A^p(\text{loc})$ can be estimated from the line width of a quadrupolar nucleus. By comparison with the observed chemical shift, $\sigma_A^d(\text{loc})$ can thus be found. For $^{23}Na^+$ dissolved in mixtures of tetrahydrofuran (THF) and amines a difference in $\sigma_A^d(\text{loc})$ of 11 ppm, with respect to $^{23}Na(H_2O)_6^+$, is reported (Delville *et al.*, 1981). As shown by the data in Table II, this is well within the 5% limit of variation in $\sigma_A^d(\text{loc})$, which is the maximum normally obtained from semiempirical SOS shielding calculations employing Eq. (6).

As demonstrated by Eqs. (10) and (12), nuclear shielding shows an inverse dependence on ΔE. This may be the controlling shielding variation influence in a series of closely related molecules where $\langle r^{-3} \rangle_{np}$, $\langle r^{-3} \rangle_{nd}$, P_μ, and D_μ are considered either to be constant or to produce mutually canceling variations. Such a case appears to arise for the ^{59}Co shielding of some diamagnetic six-coordinate cobalt(III) complexes (Kidd, 1980).

An increase in ^{59}Co shielding follows ligand replacement by a higher member of the spectrochemical series. Thus an increasing ability to cause d-orbital splitting results in a cobalt shielding increase. In another instance a correlation between the ^{59}Co chemical shift and the reciprocal of the lowest $d \rightarrow d$ transition energy is found for some C-, N-, O-, and F-bonded ligands (Freeman et al., 1957). The existence of such a correlation implies that the lowest energy transition makes the dominant contribution to ΔE and that any variation in the other shielding factors found in Eq. (12) is insignificant for the series of molecules considered. Hence such correlations are not expected to be too widely spread.

The core electron binding energy obtainable from X-ray photoelectron spectroscopy reflects the atomic charge. Thus by comparison with Eq. (6) it is anticipated that the binding energy will be related to $\sigma_A^d(\text{loc})$. For a ground state potential model of carbon nuclei Eq. (23) is found to be applicable (Jallali-Heravi et al., 1979):

$$\Delta\sigma_A^d(\text{loc}) = -0.652 \, \Delta E_b \tag{23}$$

where $\Delta\sigma_A^d(\text{loc})$ is the change in $\sigma_A^d(\text{loc})$ for a series of molecules with respect to a standard value and E_b the corresponding difference in core electron binding energies.

Although $\sigma_A^d(\text{loc})$ is a minor contributor to chemical shift differences, a change of over 7 ppm is both observed by X-ray photoelectron spectroscopy and calculated from Eq. (6) for the ^{13}C nuclei in CH_4 and CF_4. For nitrogen nuclei a range of about 13 ppm is reported for $\sigma_A^d(\text{loc})$. For a detailed comparison of ΔE_b and $\Delta\sigma_A^d(\text{loc})$ a correction due to relaxation processes should be included.

Because of the limited range of variation in $\sigma_A^d(\text{loc})$ for a given type of nucleus it seems unlikely that such changes are important in a consideration of the shielding of heavy nuclei. In addition, the dominant effect of $\sigma_A^p(\text{loc})$ ensures that no general correlation is to be expected between core electron binding energy and chemical shift.

III. Intermolecular Factors

For a pure dilute gas Eq. (16) shows that intermolecular shielding effects are density-dependent. The term $\sigma_1(T)$ in Eq. (16) is often referred to as the second virial coefficient of nuclear shielding (Jameson, 1981). In general the magnitude of σ_1 reflects the range of chemical shifts of the nucleus concerned. Thus heavier nuclei are likely to have larger values of σ_1 than lighter nuclei. In addition, nuclei on the molecular periphery usually have larger σ_1 values than those in the molecular skeleton. Invariably σ_1 is a negative quantity, such that the nuclear shielding decreases as the gas density increases. In the case of xenon gas the ^{129}Xe shielding decreases by approximately 100 ppm as the density increases by about 200 amagat, where 1 amagat = 4.4738×10^{-5} mol cm^{-3} (Jameson *et al.*, 1970).

The shielding of ^{129}Xe is found to be extremely sensitive to its physical environment. For trapped xenon atoms in the cages of the β-quinol clathrate both the shielding and its anisotropy depend on the nature of the trapping site (Ripmeester, 1982). A deshielding by up to 279 ppm is observed for a trapped xenon atom, with respect to xenon gas at zero density.

In solution the interactions between solute and solvent molecules may be considered to be either specific or nonspecific. Specific interactions include hydrogen bonding, complex formation, and those that take place with added shift and relaxation reagents. In general such interactions are predictable from a knowledge of the chemistry of the system concerned. Calculations of nuclear shielding in the presence of specific interactions may be performed using the supermolecule approach (Webb and Witanowski, 1983).

Nonspecific solvent effects on NMR parameters may be discussed on the basis of two types of models. In one the solvent is taken as a continuum and the induced shifts are correlated with a bulk property, such as dielectric constant or refractive index. The other type of model permits the shifts to be interpreted in terms of interactions between pairs of molecules.

A general pair interaction model includes terms due to van der Waals forces, bulk susceptibility effects, electric moment effects, and the magnetic anisotropy of nonspherical solutes (Raynes *et al.*, 1961).

Precise mathematical expressions are not available for many nonbonded interactions. Thus continuum models tend to be more popular for determining solvent effects on NMR parameters. Perhaps the most widely used to date is the solvaton model (Ando and Webb, 1981). By means of this model the solvent-induced shift is proportional to $(\varepsilon - 1)/2\varepsilon$, where ε is the dielectric constant of the medium.

Sum-over-states shielding calculations with INDO/S parameters, incorporating the solvaton model, predict that the nitrogen shielding of nitromethane will decrease by ~ 10 ppm as ε increases from ~ 2 to ~ 46 (Witanowski *et al.*, 1981). This is in good agreement with observation and suggests that the solvaton model may be suitable for a consideration of some of the larger shielding variations, as a function of solvent, observed for heavier nuclei such as ^{129}Xe (Schrobilgen *et al.*, 1978).

In conclusion, it has been the aim of this chapter to indicate the major factors that contribute to the observed chemical shifts of heavy nuclei and possible ways of estimating them. Perhaps the main point arising from this discussion is that the various intra- and intermolecular factors that may be operative call for extreme care when comparing various sets of experimental chemical shifts for heavy nuclei.

Acknowledgments

I am grateful to the Department of Chemistry, University of Hobart, Tasmania, for hospitality during the course of which most of this chapter was written.

References

Ando, I., and Webb, G. A. (1981). *Org. Magn. Reson.* **15**, 111.
Ando, I., and Webb, G. A. (1983). "Theory of NMR Parameters," Academic Press, London.
Ashe, A. J. (1978). *Acc. Chem. Res.* **11**, 153.
Becker, K. D. (1978). *J. Chem. Phys.* **68**, 3785.
Bendall, M. R., and Doddrell, D. M. (1978). *Aust. J. Chem.* **31**, 1141.
Buckingham, A. D., and Malm, S. M. (1971). *Mol. Phys.* **22**, 1127.
Cheremisin, A. A., and Schastnev, P. V. (1980). *J. Magn. Reson.* **40**, 459.
Delville, A., Detellier, C., Gerstmans, A., and Laszlo, P. (1981). *J. Magn. Reson.* **42**, 14.
Deverell, C. (1969). *Mol. Phys.* **16**, 491.
Ditchfield, R. (1972). *J. Chem. Phys.* **56**, 5688.
Ditchfield, R. (1974). *Mol. Phys.* **27**, 789.
Dobosh, P. A., Ellis, P. D., and Chou, Y. C. (1979). *J. Magn. Reson.* **36**, 439.
Ebraheem, K. A. K., and Webb, G. A. (1978). *Prog. NMR Spectrosc.* **11**, 149.
Ebraheem, K. A. K., Webb, G. A., and Witanowski, M. (1976). *Org. Magn. Reson.* **8**, 317.
Ellis, P. D., Chou, Y. C., and Dobosh, P. A. (1980). *J. Magn. Reson.* **39**, 529.
Freeman, R., Murray, G. R., and Richards, R. E. (1957). *Proc. R. Soc. London* **A242**, 455.
Garber, A. R., Ellis, P. D. Seidman, K., and Schade, K. (1979). *J. Magn. Reson.* **34**, 1.
Giessner-Prettre, C., and Pullman, B. (1982). *J. Am. Chem. Soc.* **104**, 70.
Gilles, D. G., Blaauw, L. P., Hays, G. R., Huis, R., and Clague, A. D. H. (1981). *J. Magn. Reson.* **42**, 420.
Golding, R. M., and Stubbs, L. C. (1979). *J. Magn. Reson.* **33**, 627.
Golding, R. M., and Stubbs, L. C. (1980). *J. Magn. Reson.* **40**, 115.
Grishin, Y. K., Sergeyev, N. M., and Ustynyuk, Y. A. (1971). *Mol. Phys.* **22**, 711.
Harris, R. K., and Mann, B. E. (1978). "NMR and the Periodic Table." Academic Press, London.
Hays, G., Gilles, D. G., Blaauw, L. P., and Clague, A. D. H. (1981). *J. Magn. Reson.* **45**, 102.
Hinton, J. F., Metz, K. R., and Briggs, R. W. (1982). "Annual Reports on NMR Spectroscopy" (G. A. Webb, ed.), (Vol. 13 p. 211). Academic Press, London.
Höller, R., and Lischka, H. (1980). *Mol. Phys.* **41**, 1041.
Ishmail, I. M., Kerrison, S. J. S., and Sadler, P. J. (1980). *J. Chem. Soc. Chem. Comm.* p. 1175.
Jallali-Heravi, M., and Webb, G. A. (1978). *J. Magn. Reson.* **32**, 429.

Jallali-Heravi, M., Webb, G. A., and Witanowski, M. (1979). *Org. Magn. Reson.* **12**, 274.
Jameson, A. K., Jameson, C. J., and Gutowsky, H. S. (1970). *J. Chem. Phys.* **53**, 2310.
Jameson, C. J. (1981). *Bull. Magn. Reson.* **3**, 3.
Jameson, C. J., and Gutowsky, H.S. (1964). *J. Chem. Phys.* **49**, 1714.
Jameson, C. J., and Jameson, A. K. (1978). *J. Chem. Phys.* **68**, 615.
Johnson, J., Hunt, J. P., and Dodgen, H. W. (1969). *J. Chem. Phys.* **51**, 4493.
Jokisaari, J., Räis, K., Lajunen, L., Passoja, A., and Pyykkö, P. (1978). *J. Magn. Reson.* **31**, 121.
Juranic, N. (1981). *J. Chem. Phys.* **74**, 3690.
Keller, H. J., and Rupp, H. H. (1971). *Z. Naturforsch.* **26a**, 785.
Kidd, R. G. (1980). "Annual Reports on NMR Spectroscopy" (G. A. Webb, ed.), Vol. 10A, p. 1. Academic Press, London.
Laitem, L., Christiaens, L., and Renson, M. (1980). *Org. Magn. Reson.* **13**, 319.
Malli, G., and Froese, C. (1967). *Int. J. Quantum Chem.* **1S**, 95.
Morishima, I., Endo, K., and Yonezawa, T. (1973). *J. Chem. Phys.* **59**, 3356.
Na-Lamphun, B., and Webb, G. A. (1983). *J. Mol. Struct.* (in press.)
Orrell, K. G. (1979). "Annual Reports on NMR Spectroscopy" (G. A. Webb, ed.), Vol. 9, p. 1. Academic Press, London.
Pople, J. A. (1957). *Proc. R. Soc. London* **A239**, 541.
Pople, J. A. (1962a). *J. Chem. Phys.* **37**, 53.
Pople, J. A. (1962b). *J. Chem. Phys.* **37**, 60.
Pople, J. A. (1964). *Mol. Phys.* **7**, 301.
Pople, J. A., and Beveridge, D. L. (1970). "Approximate Molecular Orbital Theory." Mc-Graw-Hill, New York.
Pyykkö, P. (1978). *Adv. Quantum Chem.* **11**, 353.
Ramsey, N. F. (1950). *Phys. Rev.* **78**, 689.
Ramsey, N. F. (1952). *Phys. Rev.* **87**, 1073.
Raynes, W. T., Buckingham, A. D., and Bernstein, H. J. (1961). *J. Chem. Phys.* **36**, 3481.
Ribas Prado, F., Giessner-Prettre, C., Pullman, A., Hinton, J. F., Harpool, D., and Metz, K. R. (1981a). *Theoret. Chim. Acta* **59**, 55.
Ribas Prado, F., Giessner-Prettre, C., and Pullman, B. (1981b). *Org. Magn. Reson.* **16**, 103.
Ripmeester, J. (1982). *J. Am. Chem. Soc.* **104**, 289.
Saika, A., and Slichter, C. P. (1954). *J. Chem. Phys.* **22**, 26.
Schmidt, H., and Rehder, D. (1980). *Trans. Met. Chem.* **5**, 214.
Schrobilgen, G. J., Holloway, J. H., Granger, P., and Brevard, C. (1978). *Inorg. Chem.* **17**, 980.
Stothers, J. B. (1972). Organic Chemistry (A. T. Blomquist and H. Wasserman, eds.), Vol. 24. Academic Press, New York.
Webb, G. A. (1975). "Annual Reports on NMR Spectroscopy" (E. F. Mooney, ed.), Vol. 6A, p. 1. Academic Press, London.
Webb, G. A. (1978). "N.M.R. and the Periodic Table" (R. K. Harris and B. E. Mann, eds.), p. 49. Academic Press, London.
Webb, G. A., and Witanowski, M. (1983). *Molecular Interactions 5.* (In press)
Whiffen, D. H. (1964). *J. Chim. Phys.* 1589.
Witanowski, M., Stefaniak, L. Na-Lamphun, B., and Webb, G. A. (1981). *Org. Magn. Reson.* **16**, 57.

5 Quadrupolar and Other Types of Relaxation

R. Garth Kidd

Department of Chemistry
The University of Western Ontario
London, Canada

I. Introduction

Relaxation is the process in a spectroscopic experiment whereby the Boltzmann distribution among microstates is reestablished after the equilibrium has been perturbed by resonance absorption of radiation. It is a

103

first-order process which can be characterized either by a rate constant or by its reciprocal, a relaxation time. Relaxation rates by different mechanisms are additive, and use of this parameter facilitates discussion. The reporting of relaxation data in the literature, however, is generally done in the form of relaxation times.

Optical and other high-frequency spectroscopies are not in general concerned with the relaxation process. The reason for this state of bliss lies in the Einstein theory of transition probabilities (Einstein, 1917), according to which the probability of spontaneous emission is proportional to the cube of the frequency, whereas the probability of induced absorption is proportional to the radiation density and independent of frequency. At rf's relaxation is a much slower process than at optical frequencies, and although relaxation times in optical spectroscopy are typically on the order of 10^{-8} s, the shortest NMR relaxation times are on the order of 10^{-6} s and relaxation times in the range 1–10 s are common.

The study of relaxation times per se occupies a growing number of spectroscopists, but there are two reasons why a rudimentary understanding of nuclear relaxation is essential to all who measure or interpret NMR spectral data. The first few generations of commercial spectrometers were of the continuous-wave variety, and provided care is exercised to avoid instrumental broadening, line widths provide a direct measure of relaxation times. The impact of relaxation upon spectrometer operation and the analysis of spectrometer output to provide information about molecules are both dealt with in Bloembergen's classic treatise on the subject (Bloembergen, 1961). In recent years the swing from continuous-wave to pulsed spectrometers and the sensitivity enhancement achievable after Fourier transformation place heavy demands upon both the instrument and the operator's insight when nuclei with short relaxation times are under investigation. The experimental difficulties associated with the generation of high-intensity pulses sufficiently short to give uniform power distribution over the spectrum are discussed in Section II.

In the first part of this chapter two orthogonal types of nuclear relaxation are defined, described, and related to the spectra for molecules in the liquid phase. All mechanisms whereby the energy transfer essential to the relaxation process is facilitated have two features in common, and these are laid out in broad perspective. In Sections III and IV each of these features is amplified when we focus in turn on rotational correlation times and on the energy constants coupling the spin system to the lattice. In keeping with the overall emphasis of this volume, Section V gives a broad overview of the three classes of quadrupolar nuclei and the relaxation characteristics to be expected for each.

II. Relaxation in the Region of
Motional Narrowing

Most NMR spectra are measured in the liquid state where molecules rotate with a frequency proportional to the thermal energy kT and inversely proportional to the viscous drag which increases with molecular size. For molecules of modest size in solutions not too viscous, these frequencies lie in the range 10^9-10^{12} Hz. In this region an increase in rotational frequency, as occurs on heating the sample or moving to a solvent of lower viscosity, causes a decrease in relaxation rate, which results in a narrowing of the spectral line. Within this relaxation region, known as the motional narrowing region, any increase in either the rate of molecular rotation or the spectroscopic irradiation frequency results in resonance line narrowing. In solid samples or highly viscous liquids where molecular reorientation is restricted and rotational frequencies are low relative to the irradiation frequency, motional narrowing does not obtain, and an increase in rotation frequency can result in an *increase* in the relaxation rate. This relationship is illustrated in Fig. 1 where the curves for the two different relaxation rates diverge beyond the region of motional narrowing. It is also evident that within the region of motional narrowing the rates for the two distinct relaxation processes are the same, and only outside this region does one consider for the principal relaxation mechanisms factors that might differentiate R_1 from R_2.

A. Spin–Lattice (T_1) and Spin–Spin (T_2) Relaxation

The cause–effect relationship in an NMR experiment has been pictured in two different ways, each of which gives rise to a pair of definitions for the two relaxation times. The phenomenological view arises out of classical electromagnetic theory based upon Ampere's law, Gauss's law, and Maxwell's equations and is illustrated in Fig. 2a. This figure shows an ensemble of identical nuclei precessing randomly about the external field direction such that the macroscopic magnetization vector given by the ensemble vector sum lies coincident with B_0. Application of a rf field satisfying the resonance condition introduces phase coherence to the Larmor precession cone and tilts the magnetization vector away from the external field in the z-direction, giving it a transverse component in the xy plane. Following removal of the rf field, the transverse component decays to zero according to a first-order rate law as the magnetization vector resumes coincidence with the z axis. The process is designated

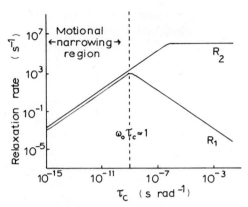

Fig. 1 The rates for longitudinal (R_1) and transverse (R_2) relaxation as a function of rotational correlation time.

transverse relaxation and is characterized by a rate constant. $R_2 = 1/T_2$, where T_2 is known as the transverse relaxation time. Thermodynamically the spin ensemble is out of equilibrium and the process represents an *entropy* increase for the system. The loss of phase coherence by the precessing nuclei is also referred to as spin–spin relaxation. If the external field B_0 is switched off while the nuclear magnetization vector is coincident with the z axis, this vector decays exponentially to zero as the z-components of the individual spin vectors are randomized. The rate constant governing this process is $R_1 = 1/T_1$, where T_1 is known as the

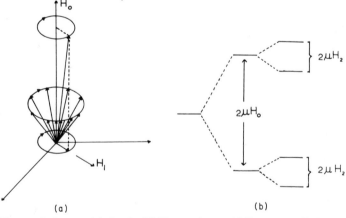

Fig. 2 Two equivalent models for the NMR experiment. (a) Torques applied to precessing nuclear magnets. (b) Magnetic perturbations of nuclear spin states.

longitudinal relaxation time. This relaxation process requires the transfer of *energy* from the nuclear ensemble to the surrounding electrons and for this reason is also referred to as the spin–lattice relaxation time. In solids where molecular motion is restricted and the spin system is insulated from the lattice, R_1 can be several orders of magnitude less than R_2, and spin–lattice relaxation times of several hours can be achieved.

The quantum mechanical view of an NMR experiment is based upon time-dependent perturbation theory and is illustrated in Fig. 2b. Here the energy of a nuclear spin state relative to its zero-field value is depicted as a function of two discrete perturbations, the largest resulting from application of the external magnetic field B_0 and the smaller, time-dependent perturbation resulting from the rf field B_1. Viewed according to this model, T_1 is the relaxation time governing transitions from states of the upper manifold to states of the lower manifold and T_2 is the relaxation time governing transitions from the upper to the lower state within a manifold. Spectroscopic line widths are inversely proportional to the life-times of the energy states under observation according to the Heisenberg uncertainty principle, and because this principle is a construct of quantum mechanics having no counterpart in classical physics, the dependence of line width upon relaxation time is seen most readily in Fig. 2b. Spin–lattice relaxation from an upper to a lower manifold state establishes an upper limit for the lifetime of the excited state and a lower limit for the line width for the transition. Relaxation within a manifold, the so-called spin exchange phenomenon, can serve to further limit the lifetime of a spin state, and in these instances the spin–spin relaxation time T_2 obtainable from a measured line width is less than T_1. In circumstances where T_2 is not equal to T_1, it may be shorter than T_1, but clearly it can never be longer than T_1, as is reflected in Fig. 1.

B. Mechanisms for Energy Divestiture

In samples of NMR interest the spin system is well insulated from the lattice, and in the most thermally isolated instances special means for accelerating reestablishment of the off-resonance Boltzmann distribution must be adopted. This results from nuclei being buried deep within molecules. Such a situation contrasts sharply with other forms of absorption spectroscopy where saturation is not a problem. This difficulty has been turned to advantage over the past 30 years, so that today measured values of relaxation times can provide us with such important parameters as nuclear quadrupole coupling constants, electric field gradients at specific locations within a molecule, nuclear quadrupole moments, spin–rotation coupling constants, values for unresolved scalar coupling constants to quadrupolar nuclei, interatomic distances, and rates of molecular rotation. The connection between each of these molecular properties and the measured relaxation time is provided by a theoretical model describing how a nucleus divests itself of energy (or acquires entropy) in the relaxation process.

The nucleus may couple with its surroundings either magnetically or electrically. Instances of both are identified in the detailed consideration of coupling mechanisms presented in Section IV, and in every case the strength of the coupling is represented by a coupling constant. In promoting relaxation, coupling is a necessary but not a sufficient condition. Loss of energy by the spin system can be regarded as a stimulated emission process, and the probability of stimulated emission involves a frequency dependence. Frequencies arise in the coupling of the nucleus to its surroundings through time modulation of the coupling constant by rotation of some part or all of the molecule. A characteristic correlation *time* is the form in which this modulation enters the model for each relaxation mechanism; the connection between molecular rotation and correlation times, along with the molecular properties that influence correlation times, are presented in Section III.

The two factors present in every relaxation mechanism are a coupling constant A and a correlation time τ. Since the overall rate of relaxation has units of reciprocal seconds, the coupling constant must appear as a squared term, and the general equation for each mechanism is

$$R(s^{-1}) = \text{const} \times A^2(Hz^2)\tau(s) \tag{1}$$

where parenthetical symbols represent the units for the respective variables.

When considering relaxation data from the literature, the reader should be acutely aware of a potential pitfall related to the units in which coupling constants are measured. A frequency quoted simply in units of reciprocal seconds is not fully specified, because this can represent either cycles per second (hertz) or radians per second, and because the coupling constant appears as a squared term in the relaxation equation, the difference is a factor of $(2\pi)^2 = 39$. Because this factor is frequently encountered in the energy–frequency conversion where its presence or absence is designated simply by the bar through Planck's constant h, another potentially rich source of confusion should be recognized. As in other areas of science where subtle distinctions are crucial, the principle of *caveat lector* must apply.

III. The Rotational Correlation Time

Rotational correlation describes for molecules in motion their *orientation* with respect to one another compared with their relative orientation at an earlier time. Nuclear relaxation is a cooperative phenomenon which must necessarily be described using parameters characteristic of a molecular *ensemble*. The temporal characteristics of an ensemble are described by the time periods in which various properties remain correlated in the face of ceaseless random particle motions which cause the rapid decay of any particular instantaneous arrangement perceivable using a "fast-shutter" spectroscopic technique. The time-dependent molecular orientation is represented mathematically by an autocorrelation function $F(t)$, whose value goes from 1 to 0 as the relative orientations go from the correlated to the randomized states. The time constant that characterizes this exponential decay is defined as the rotational correlation time τ_c.

Defined in this way, the correlation time τ_c is an ensemble property, and it is this correlation time that enters into theoretical analyses of nuclear relaxation. Closely related to but not identical with τ_c is the single-molecule time constant τ_θ defined as the time period for 1 rad of molecular rotation (Huntress, 1970). Since most chemists who measure relaxation times are motivated by a desire to characterize individual molecules rather than molecular ensembles, it is the τ_θ interpretation of rotational correlation that is frequently used in discussions of relaxation mechanisms.

A. The Correlation Function in the Time Domain

In the classic study on NMR that appeared as his Ph.D. thesis in 1948, Bloembergen (1961) first showed that the energy transfer essential to relaxation will occur if the position vectors determining instantaneous magnetic dipolar or electric quadrupolar coupling of a nucleus with its surroundings are functions of time. In a liquid, these functions $F(t)$ vary with time in a random fashion as the molecules containing the magnetic nuclei undergo Brownian motion. Although the value of $F(t)$, being a measure of magnetic or electric coupling, is nonzero, it is the *differences* between $F(t)$ values over short intervals of time τ that are characterized by the *correlation function* $k(\tau)$.

$$k(\tau) = F(t)F^*(t + \tau) \qquad (2)$$

This function represents the decay in correlation as the time interval τ becomes longer. For small values of τ, F^* at $t + \tau$ is highly correlated with

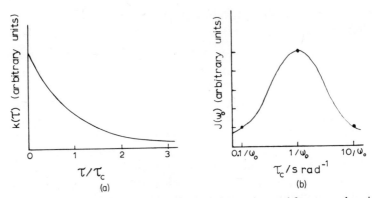

Fig. 3 Representations of the correlation function in the time and frequency domains. (a) Exponential decay of correlation function defining rotational correlation time τ_c. (b) Spectral density of the ω_0 Fourier component for ensembles with different τ_c values.

the initial value of F, whereas for large τ, F and F^* become independent variables which are no longer correlated.

For the molecular processes that presently concern us, $k(\tau)$ can be assumed to take the form

$$k(\tau) = \text{const} \times e^{-\tau/\tau_c} \tag{3}$$

as illustrated in Fig. 3a. In general, $k(\tau)$ is a function that goes rapidly to zero when τ exceeds a characteristic value τ_c known as the *correlation time*. This equation thus defines τ_c for the ensemble as the length of time required by the kT randomizing force to reduce $F(t)$ to $1/e$ or 37% of its initial value, and the exponentially decaying curve one obtains is known as the correlation function in the *time domain*.

A magnetic or electric field fluctuating at the Larmor frequency promotes relaxation, and in order to recognize the frequencies hidden within the continuously decaying $k(\tau)$ function, it is subjected to a mathematical device known as Fourier transformation which converts it from a time function into a frequency function.

B. The Spectral Density and the Frequency Domain

The thermal randomizing force driving rotational and translational motion of the molecules brings about exponential decay of a correlation function (we can readily perceive that there are *frequencies* associated with the periodic nature of rotations). Fourier transformation of the correlation function

$$J(\omega) = \int k(\tau)e^{i\omega\tau}\, d\tau \tag{4}$$

casts it into the *frequency domain* and yields a spectrum of frequencies $J(\omega)$ known as the *spectral density*, illustrated in Fig. 3b. The intensity of the fluctuation in $F(t)$ at a particular frequency is obtained by substituting Eq. (3) into Eq. (4) and yields

$$J(\omega) = \text{const} \times \tau_c/[1 + (\omega\tau_c)^2] \tag{5}$$

confirming the fact that the Fourier transform of an exponential is a Lorentzian. The independent variable ω in this equation is the label designating the individual frequency components of the Fourier spectrum. These frequencies modulate the approach to zero of the correlation function and provide an oscillating magnetic field at the nucleus, *one* component of which occurs at the Larmor frequency and thus makes a contribution to the relaxation proportional to its spectral density $J(\omega_0)$.

A representation of Eq. (5) in the frequency domain is obtained by plotting $J(\omega)$ versus ω for an ensemble with a particular τ_c. Figure 4 contains these plots for ensembles having three different correlation times. The curves are normalized to equal kinetic energy in each ensemble by maintaining equal areas, and the height of the low-frequency plateau rises as the horizontal extension of the curve is reduced by selecting ensembles with progressively longer τ_c. In each Fourier spectrum the intensity of the individual frequency components remains constant for low frequencies below τ_c^{-1}, falling off rapidly to zero at high frequencies. The half-intensity point on each curve occurs for the component whose frequency is τ_c^{-1}, and Fig. 4 shows that for the ω_0 frequency component promoting relaxation $J(\omega_0)$ is maximized in the ensemble whose τ_c equals ω_0^{-1}. Curve B represents the ensemble with $\tau_c = 10^{-8}$ s.

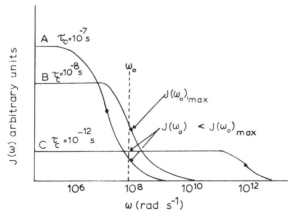

Fig. 4 Fourier spectra in the frequency domain for molecular ensembles with different correlation times.

If the nucleus undergoing relaxation in this ensemble has a Larmor frequency $\omega_0 = 10^8$ rad s^{-1}, then the spectral density at this frequency is that designated $J(\omega_0)_{max}$ in Fig. 4. Curves A and C, representing ensembles having τ_c values longer and shorter, respectively, than that represented by curve B, both have spectral densities at ω_0 that are less than $J(\omega_0)_{max}$ as shown. Although the spectral density for ensemble B is higher than $J(\omega_0)_{max}$ at frequencies lower than ω_0, the resonance nature of the phenomenon makes these higher intensities less effective in promoting relaxation.

An alternative representation of Eq. (5) is the familiar Lorentz curve in Fig. 3b. Here the intensity of the ω_0 frequency component is plotted as a function of τ_c for the ensemble, and again we see that spectral density at the Larmor frequency is maximized when τ_c for the ensemble coincides with ω_0^{-1}. Correlation times above and below this value are both less effective. Because τ_c in the picosecond range is much shorter than the shortest Larmor period ω_0^{-1} experimentally achievable, changes such as viscosity increase and temperature decrease that lengthen τ_c for the ensemble increase the intensity of the spectral component with frequency ω_0, increasing the relaxation rate and thus broadening the spectral line.

C. Factors Influencing Rotational Correlation Time

With superconducting solenoid fields of 7 T, the resonance frequency for all nuclei is less than 10^{10} rad s^{-1}, whereas that for most nuclei lies in the region of 10^8 rad s^{-1} or lower. For molecules undergoing liquid state NMR observation, the rotation period is about 10 ps rad^{-1}, putting τ_c^{-1} in the region of 10^{11} rad s^{-1}, three orders of magnitude greater than the resonance frequency. Hence, relaxation is relatively inefficient. The molecular ensembles present in liquids are represented by curve C in Fig. 4 and have a spectral density $J(\omega_0)$ which would increase in magnitude and enhance the relaxation rate if the height of the curve-C plateau were raised to $J(\omega_0)_{max}$ by slowing the rate of molecular rotation, thereby lengthening τ_c.

1. Viscosity

Rotational correlation time is directly proportional to solution viscosity, consistent with a dynamic model in which the rate of molecular rotation is limited by viscous drag. By selecting chemically inert solvents that span a range of viscosities, it is possible to vary τ_c by a factor of 10 or more (Wasylishen et al., 1977). Increases in solvent viscosity lengthen τ_c and enhance $J(\omega_0)$, thereby increasing the relaxation rate and broadening the spectral line.

2. Temperature

By virtue of supplying rotational kinetic energy to the molecules, a temperature increase causes more rapid rotation and reduces τ_c. In addition, viscosity also decreases, giving rise to a further reduction in τ_c. With both factors operating in the same direction, temperature effects on τ_c are substantial, and significant decreases in line width with increasing temperature are expected. Where line widths are found to *increase* with increasing temperature, one of three possible causes should be considered: (a) chemical exchange between or among magnetically inequivalent sites, (b) relaxation primarily by the spin–rotation mechanism (Section IV,B), or (c) relaxation primarily by the scalar coupling mechanism in the Zeeman-separated case (Section IV,E).

3. Molecular Size

The Stokes–Einstein–Debye (SED) model (Debye, 1929; Edward, 1970) for molecular rotation contains a direct proportionality between rotation period and molecular volume. For molecules approximated as spherical, τ_c is proportional to the cube of the molecular radius and can be obtained by scaling from the τ_c for a molecule of known radius. Attempts to evaluate a proportionality constant that is consistent from one system to another have been only partially successful. Semiempirical corrections are usually incorporated into the model in the form of microviscosity factors (Gierer and Wirtz, 1953; Kivelson *et al.*, 1969).

4. Molecular Shape

Molecules represented as spheres can rotate without solvent displacement, and in the slip limit can rotate with zero viscous drag. Ellipsoidal or spheroidal molecules must displace solvent as they rotate, inhibiting molecular rotation by internal friction of solvent displacement and making τ_c proportional to the volume of solvent displaced. Models for calculating rotational correlation times of ellipsoidal (Hu and Zwanzig, 1974) and spheroidal (Youngren and Acrivos, 1975) molecules in the slip limit are available, and a critical analysis of their reliability has been carried out (Boere and Kidd, 1983).

IV. Relaxation Interactions Modulated by Molecular Rotation

The selection rules that govern transitions between nuclear spin states are such that the probability of absorption is equal to the probability of stimulated emission, and the restoration of a Boltzmann distribution among spin states following resonance absorption of energy is dependent upon the spontaneous emission process. The nuclear species that come under NMR investigation differ from one another in terms of magnetic moment, nuclear quadrupole moment, and bonding radius of the atom in which the nucleus is located. These values, together with the electric field gradient at the nucleus under study, determine the relative magnitudes of the mechanisms contributing to relaxation.

Where more than one mechanism contributes to the overall relaxation process, experimental techniques have been developed for fractionating the total relaxation by mechanism. Because each of the five dominant mechanisms depends strongly upon at least one of the four variables previously identified, one can anticipate which mechanism is likely to carry most of the relaxation in any particular instance, and this knowledge is extremely useful when making spectral assignments in a complicated spectrum. The order of presentation is from the most to the least effective in molecules where the mechanism in question is favored, and Table I provides a quantitative summary of the relationships.

A. Nuclear Quadrupole Coupling

Coupling between a nuclear electric quadrupole moment Q and an electric field gradient eq generated by the nuclear environment can provide the most effective route for spin–lattice relaxation. Quadrupolar relaxation rates as high as 10^6 s^{-1} are known, and for nuclei such as ^{181}Ta with quadrupole moments of about 3 barns, relaxation limits the lifetime of the excited state to such an extent that uncertainty broadening makes the NMR signal unobservable even in highly symmetric environments having minimum field gradients (Erich et al., 1973).

The rate for quadrupole relaxation in the motional narrowing limit is given by the equation (Abragam, 1961)

$$R^q = \frac{1}{T^q} = \frac{3\pi^2}{10} \frac{2I + 3}{I^2(2I - 1)} \left(\frac{e^2qQ}{h}\right)^2 \left(1 + \frac{\eta^2}{3}\right) \tau_\theta \qquad (6)$$

where, in addition to the nuclear quadrupole coupling constant (NQCC) and the rotational correlation time terms, we find a numerical coefficient that varies from 14.8 to 0.112 with variation in spin number I and a factor containing η that accounts for the asymmetry of the electric field gradient. Values of the numerical coefficient and other parameters in the quadrupolar relaxation equation for specific nuclei are given in Table II. Some authors (Sharp, 1972; Saito et al., 1973; Jacobsen and Schaurenberg,

TABLE I
Summary of Relaxation Mechanisms

Mechanism	Interaction	Maximum R (s^{-1})	Maximum coupling constant range	Correlation time (s)
Quadrupolar	Electric	10^6	0.24 MHz (^{2}H) to 150 MHz (^{35}Cl)	τ_θ typically 10^{-12}–10^{-10}
Spin–Rotation	Magnetic	10^2	0.5 kHz (^{13}CCl) to 116 kHz (^{31}PH)	τ_θ typically 10^{-13}–10^{-11}
Dipole–dipole	Magnetic	1	33 kHz (^{195}PtH) to 140 kHz (^{13}CH)	τ_θ typically 10^{-12}–10^{-10}
Shielding anisotropy	Magnetic	10^{-1}	$\sigma_\parallel - \sigma_\perp = 1600$ ppm minimum for 0.16 MHz and $R^{sa} = 0.05$ s^{-1} $\sigma_\parallel - \sigma_\perp \approx 5000$ ppm gives 0.5 MHz for ^{199}Hg	τ_θ typically 10^{-12}–10^{-10}
Scalar coupling	Magnetic	10^{-1} $R_2^{sc} < 10^3$	$J_{IS} < 10^4$ Hz	τ_s typically $> 10^{-6}$

TABLE II

Constants and Molecular Parameters Determining Relaxation Rates for Quadrupolar Nuclei

Nucleus	I	Q (10^{-28} m^2)	$\left(\dfrac{3\pi^2}{10}\right)\dfrac{2I+3}{I^2(2I-1)}$	Maximum NQCC (MHz)	Maximum R^Q for $\tau = 10$ ps (s^{-1})	Minimum T (ms)
^{2}H	1	2.7×10^{-3}	14.8	0.24	8	100
^{6}Li	1	6.9×10^{-4}	14.8	0.01	0.02	5×10^4
^{9}Be	$\frac{3}{2}$	5.2×10^{-2}	3.95	0.35	5	200
^{10}B	3	7.4×10^{-2}	0.592	6.2	230	4
^{11}B	$\frac{3}{2}$	3.6×10^{-2}	3.95	3.0	360	3
^{14}N	1	1.6×10^{-2}	14.8	6.0	5300	0.2
^{17}O	$\frac{5}{2}$	-2.6×10^{-2}	0.947	14.0	2000	0.5
^{23}Na	$\frac{3}{2}$	1.4×10^{-1}	3.95	2.1	175	5
^{25}Mg	$\frac{5}{2}$	2.2×10^{-1}	0.947	1	10	100
^{27}Al	$\frac{5}{2}$	1.5×10^{-1}	0.947	7	500	2
^{33}S	$\frac{3}{2}$	-6.4×10^{-2}	3.95	15	9000	0.1
^{35}Cl	$\frac{3}{2}$	-7.9×10^{-2}	3.95	150	9×10^5	1×10^{-3}

1977), by incorporating $\hbar$ rather than h into the denominator, express the NQCC in angular frequency units of radians per second rather than the customary hertz. In these instances, the NQCC is numerically greater by a factor of 2π, and the numerical coefficient becomes $\frac{3}{40}$ in compensation.

The range of NQCCs observed for quadrupolar nuclei in typical covalently bonded environments is given in Table II. When combined with a representative τ_θ of 10 ps, they yield maximum quadrupolar relaxation rates up to 10^6 s^{-1} for ^{35}Cl, whereas for ^{2}H with its lower quadrupole moment, the maximum rate is about 10 s^{-1}. This is still an order of magnitude greater than the fastest relaxation achievable by the dipole–dipole mechanism and two orders of magnitude faster than that achieved through shielding anisotropy. Only the spin–rotation mechanism, which may achieve rates up to 10^2 s^{-1}, can compete with the quadrupolar mechanism and then only in instances where the NQCC is less than about 3 MHz either because of a low quadrupole moment or a low electric field gradient. In all other cases, the quadrupole mechanism dominates to such an extent that the contribution from any other mechanism is immeasurably small.

B. Spin–Rotation Coupling

Molecules in which the electric charge distribution is not spherically symmetric possess a *molecular* magnetic moment under rotation of this asymmetric electronic charge distribution. The interaction between a nuclear magnetic moment and the molecular magnetic field produced at the position of the nucleus is designated the spin–rotation coupling constant C and is measured in hertz. The strength of this interaction is typically three orders of magnitude less than the quadrupole coupling constants previously discussed. Modulation of the interaction by changes in the *rate* of molecular rotation causes nuclear magnetic relaxation at a rate given by (Abragam, 1961)

$$R^{sr} = 1/T^{sr} = (2IkT/\hbar^2)C^2\tau_J \qquad (7)$$

A freely tumbling molecule whose rate of rotation depends only on its moment of inertia I and the absolute temperature T rotates through an angle of θ radians in $\tau_\theta = \theta(I/3kT)^{1/2}$ seconds and is said to be in the *inertial region* of motion. In liquids, molecular reorientation is impeded by a viscosity-related frictional force transforming at the molecular level into collisions having varying degrees of "stickiness." This *diffusion model* represents molecules undergoing at each collision a small random jump characterized by a rotational diffusion tensor defined in such a way that the molecule rotates or diffuses through an rms angle of θ radians in τ_θ seconds. In the diffusion model, the *velocity* or rotation changes only on collision, and a time period $\tau_J < \tau_\theta$ is formally identified with the average time between collisions. Although the jump nature of the motion makes it incremental, it is regarded as continuous because many jumps are required to achieve a diffusion angle of 1 rad. Thus both an *angular* time constant τ_θ and a *velocity* (or angular *momentum*) time constant τ_J are associated with the diffusion picture of molecular rotation. Unlike the time constant appearing in the relaxation equations for other mechanisms, it is the *velocity* correlation time that appears in Eq. (7).

In most solution studies, the molecular motion is a diffusion process, and under these circumstances τ_J and τ_θ are inversely related to one another through the Hubbard relationship (Hubbard, 1963) $\tau_\theta\tau_J = I/6kT$, so that Eq. (7) becomes

$$R^{sr} = (2I^2/6\hbar^2)(C^2/\tau_\theta) \tag{8}$$

Relaxation occurring predominantly by the spin–rotation mechanism is identified by a relaxation rate, hence a line width, that *increases* with increasing temperature, as implied by Eq. (8). Quadrupolar and dipolar relaxation rates *decrease* with increasing temperature because their equations contain τ_θ in the numerator rather than in the denominator.

The Table I appearance of this relaxation mechanism in second place should not be interpreted as suggesting a widespread incidence of spin–rotation relaxation. The maximum room temperature rates observed are less than 10 s^{-1}, and the maximum rate of 10^2 s^{-1} appearing in Table I is achieved only at elevated temperatures because of the inverse temperature dependence of this particular relaxation process. The heavier spin-$\frac{1}{2}$ nuclei such as ^{119}Sn, ^{125}Te, ^{199}Hg, and ^{207}Pb are now beginning to come under regular NMR observation, however, and in many compounds their relaxation is dominated by the spin–rotation mechanism, because covalent radius factors minimize dipole–dipole contributions. Definitive relaxation studies where significant spin–rotation contributions might be expected are summarized in Table III. It is worth noting that the approximately inverse relationship existing between molecular moment of inertia and spin–rotation coupling constant places a natural limit on the spin–rotation relaxation rate that can be achieved, regardless of molecular structure.

The spin–rotation relaxation rate provides a link between relaxation theory and chemical shift theory that has proven extremely valuable in locating the standard states necessary for establishing absolute shielding scales. Deverall (1970) has observed that, to within 5%, the paramagnetic component of nuclear shielding is proportional to the spin–rotation cou-

TABLE III

Incidence and Scope of Spin–Rotation Relaxation

Molecule	Temperature (°C)	I (10^{-40} g cm^2)	C (kHz)	R_1^{sr} (s^{-1})	Percentage of Total Relaxation	Ref.
^{13}CH$_3$Br	5	⊥ 87.7 ∥ 5.5	-1.20 -21.2	0.048	38	Lassigne and Wells (1977a)
^{13}CF$_4$	—	—	—	0.048	—	Rugheimer and Hubbard (1963)
^{19}FClO$_3$	27	160	140	4.4	100	Maryott et al. (1971)
^{29}SiMe$_4$	25	—	—	0.052	98	Levy and Cargioli (1974)
^{31}PCl$_3$	25	434	-5.15	0.25	100	Aksnes et al. (1968)
^{119}SnCl$_4$	25	858	-6.1	0.63	100	Sharp (1972)
	150	858	-6.1	2.2	100	Sharp (1972)
^{119}SnI$_4$	150	4270	-0.66	0.25	12	Sharp (1972)
	220	4270	-0.66	3.3	100	Sharp (1972)
^{199}HgMe$_2$	-39	164	97	0.51	80	Lassigne and Wells (1977b)
	27	164	97	0.78	90	Lassigne and Wells (1977b)
^{207}PbCl$_4$	-6	928	7.8[a]	0.14	93	Hawk and Sharp (1974)
	40	928	7.8[a]	0.4	100	Hawk and Sharp (1974)

[a] Assuming J diffusion; M diffusion gives 4.5.

pling constant C. For nuclei relaxed primarily by spin–rotation, a correlation between line widths and chemical shifts should be observed, and extrapolation to zero line width yields the $\sigma_p = 0$ position on the chemical shift axis, which is otherwise obtainable only from molecular beam or optical pumping experiments.

C. Dipole–Dipole Coupling

The first nuclei to be extensively studied by NMR spectroscopy were ^{1}H and ^{19}F, and even today proton studies represent a larger fraction of the total NMR literature than those based on any other nucleus. With the growth of multiple-scan methods for signal enhancement over the past 15 yr, interest in ^{13}C spectroscopy has increased to the point where today it occupies second place. Common to all three of these nuclei are a spin $I = \frac{1}{2}$, a covalent radius less than 0.8 Å, and a relatively large nuclear magnetic moment. Taken together these three factors result in relaxation occurring predominantly or exclusively by the dipole–dipole mechanism, and chemists whose NMR experience is limited to one or the other of these nuclei can mistakenly conclude that other mechanisms are unlikely to make a significant contribution. Although receiving less attention in the past, ^{15}N also has characteristics that favor dipolar relaxation, and Fourier transform instrumentation brings this isotope within the realm of routine observation (Levy *et al.*, 1976).

The magnetic coupling between a relaxing nucleus and other magnetic nuclei in the same molecule provides an interaction whose modulation by molecular rotation causes relaxation. Where the coupling is between identical isotopes, the relaxation rate is given by (Abragam, 1961).

$$R^{dd} = 1/T^{dd} = 2\gamma_I^4\hbar^2 I(I + 1)r^{-6}\tau_\theta \tag{9}$$

where I is the spin number of the nucleus and r the dipole–dipole separation. Where the coupling is between different isotopes, the relaxation rate after differences in γ and I have been accounted for is two-thirds that in the identical isotopes case and is given by (Abragam, 1961)

$$R^{dd} = 1/T^{dd} = \tfrac{4}{3}\gamma_I^2\gamma_S^2\hbar^2 S(S + 1)r^{-6}\tau_\theta \tag{10}$$

in which S is the spin number of the other nucleus.

The r^{-6} dependence of the dipolar relaxation rate makes this an effective mechanism only when two dipoles are in close proximity, and in operational terms this means about 2 Å or less. A number of consequences limiting the incidence of dipole–dipole relaxation follow from this proximity rule: (a) Only intramolecular couplings to directly bonded atoms are generally effective. Intermolecular couplings may be significant in cases of dissolved paramagnetic impurities and of solvent protons in Van der Waals contact with other protons. In both these cases translational motion modulates the coupling and a different relaxation equation containing a translational correlation time governs. (b) The larger the covalent radius of the target atom, the less effective the dipolar mechanism, and beyond the first row of the periodic table the dipolar contribution to the total relaxation is small. In a comparison of ^{29}Si and ^{13}C, their radius ratio raised to the sixth power is a factor of 13. (c) Compared with the hydrogen covalent radius of 0.37 Å, all other bonded atoms have covalent radii that put the bond length outside the 2-Å limit and make their dipolar contribution to the relaxation minimal.

A prominent example of dipolar relaxation involves the ^{13}C atom to which protons are covalently bonded, and rates in the range 0.1–1.0 s^{-1} have been observed.

Quadrupolar relaxation rates, where present, are several orders of magnitude greater than this, and the dipolar mechanism makes a significant contribution to the relaxation process only for $I = \frac{1}{2}$ nuclei which have no quadrupole moment. In addition, the atom to which the $I = \frac{1}{2}$ nucleus is bonded must have a nonzero magnetic moment. In this context it must be recalled that fewer than 2% of all carbon and oxygen atoms have a nuclear moment. At the present time, ^{13}C and ^{15}N bearing one or more covalently bonded hydrogen atoms provide examples of dipolar relaxation whose interpretation is open to the least ambiguity. ^{29}Si and ^{31}P in certain environments derive a significant portion of their relaxation from dipolar interactions, but other mechanisms make significant contributions as well and the nuclear Overhauser enhancement (NOE) technique must be used to determine the fraction of total relaxation that is dipolar.

For $I = \frac{1}{2}$ nuclei bonded to hydrogen atoms, Eq. (10) takes the form

$$R^{dd} = n\hbar^2 \gamma_i^2 \gamma_H^2 r_{iH}^{-6} \tau_\theta \tag{11}$$

where n is the number of hydrogen atoms bonded to the nucleus under observation; molecular constants for use in Eq. (11) are given in Table IV. The fraction of the total relaxation rate that is dipolar can be identified by measuring the NOE through proton decoupling. The theoretical maximum NOE for a particular nucleus, observable when 100% of the relaxation is dipolar, is calculated from $\gamma_H/2\gamma_i$. The experimentally observed NOE_{exp} defined as (double resonance intensity/single resonance intensity) $- 1$ yields the dipolar component of the total relaxation rate from the equation

$$R^{dd} = [NOE_{exp}/(\gamma_H/2\gamma_i)]R^{total} \tag{12}$$

Experimental NOE values can routinely be measured with a precision of $\pm 15\%$, and $\pm 5\%$ can be achieved in favorable cases.

TABLE IV

Molecular Constants Determining Dipolar Relaxation Rates for Spin-$\frac{1}{2}$
Nuclei Covalently Bound to One Hydrogen

Nucleus	γ_i (10^3 rad s^{-1} G^{-1})	$\gamma_H/2\gamma_i$	r_{iH} (10^{-8} cm)	$\hbar^2\gamma_i^2\gamma_H^2 r^{-6}$ (s^{-2})
^{1}H	26.75	0.5	0.74	4.76×10^{12} [a]
^{13}C	6.73	1.988	1.107	1.94×10^{10}
^{15}N	−2.71	−4.94	1.01	5.46×10^9
^{29}Si	−5.31	−2.52	1.48	2.12×10^9
^{31}P	10.83	1.24	1.40	1.23×10^{10}
^{77}Se	5.10	2.62	1.47	2.03×10^9
^{119}Sn	−9.97	−1.34	1.70	3.25×10^9
^{195}Pt	5.75	2.32	1.7	1.1×10^9

[a] Corrected for the $\frac{3}{2}$ effect.

D. Shielding Anisotropy

The portion of the external magnetic field B_0 imposed on a molecule by
a NMR spectrometer and actually experienced by a nucleus is determined
by the nuclear shielding constant σ according to

$$B_{loc} = B_0 - \sigma B_0 = B_0(1 - \sigma) \tag{13}$$

If the electronic distribution is anisotropic, σ will vary with direction and
different chemical shifts will be observed along different molecular axes.
Because the 10^{10}–10^{12} s^{-1} rate of molecular rotation in mobile liquids is
several orders of magnitude greater than typical observation frequencies
of 10^8 s^{-1}, the individual components of the shielding tensor are mixed
and only the average value or trace of the tensor $\sigma_{av} = \frac{1}{3}(2\sigma_\perp + \sigma_\parallel)$ or σ_{av}
$= \frac{1}{3}(\sigma_{xx} + \sigma_{yy} + \sigma_{zz})$ is observed experimentally. Because the instantaneous B_{loc} is orientation-dependent because of the anisotropy in σ, however,
the magnetic coupling between B_0 and the relaxing nucleus is modulated
by the molecular rotation at frequencies determined by τ_θ. The intensities
of the frequency components that promote nuclear relaxation are again
given by functions similar to those represented in Fig. 3, and in the region
of motional narrowing where $\omega\tau_\theta < 1$, relaxation rates for symmetric top
molecules are given by (Abragam, 1961)

$$R_1^{sa} = \tfrac{2}{15}\gamma^2 H_0^2(\sigma_\parallel - \sigma_\perp)^2\tau_\theta \tag{14}$$

$$R_2^{sa} = \tfrac{7}{45}\gamma^2 H_0^2(\sigma_\parallel - \sigma_\perp)^2\tau_\theta \tag{15}$$

In the case of relaxation by shielding anisotropy, $R_1{}^{sa}$ and $R_2{}^{sa}$ are differentiated because, even in the motional narrowing region, the nature of the correlation function is such that they differ by about 15%. It should be borne in mind, however, that this difference is probably less than the experimental uncertainty in most R^{sa} determinations.

Shielding anisotropy makes a measurable contribution to relaxation if the anisotropy term $\sigma_\parallel - \sigma_\perp$ exceeds some threshold value which is achieved in practice only with atoms that experience a large range of chemical shifts. Typical values of 5×10^3 rad $s^{-1} G^{-1}$ for γ; 2.4×10^4 G for B_0, and 10^{-11} s for τ_θ require $\sigma_\parallel - \sigma_\perp$ to be at least 1600 ppm if an R^{sa} of 0.05 s^{-1} is to be achieved. The ^{199}Hg shielding in dimethylmercury has an anisotropy of 4600 ppm, and shielding anisotropy contributes 10% of the total 0.88 s^{-1} relaxation rate at 300 K. Since the remaining 90% comes from the spin–rotation mechanism, the shielding anisotropy fraction increases with decreasing temperature.

The presence of B_0 in Eqs. (14) and (15) makes the shielding anisotropy relaxation rate field or frequency-dependent. Not only does this provide an experimental criterion for identifying the presence of a shielding anisotropy contribution to the total relaxation rate, but it also means that molecules whose anisotropy is not sufficiently large to yield a measurable R^{sa} at conventional field strengths may display one at the elevated fields now becoming available.

E. Scalar Coupling

Where intramolecular spin–spin (scalar) coupling exists (Table V) the strength of this interaction is represented by the familiar constant J_{IS}, measured in hertz, and can be expressed as $A = 2\pi J$ radians per second. Relaxation of both I and S at rates less than A results in a $2nS + 1$ multiplet for nucleus I. If, as usually occurs for a quadrupole-relaxed S, the relaxation rate of S is greater than the coupling constant A, the local field produced at I by S fluctuates with a correlation time $\tau_S = T_{1(S)}$ and causes the relaxation of I at a rate proportional to the square of the coupling constant. Under these circumstances, the relaxation rates for nucleus I by this mechanism are (Abragam, 1961)

$$R_1{}^{sc} = n_S \frac{8\pi^2}{3} J_{IS}{}^2 S(S + 1) \frac{\tau_S}{1 + (\omega_I - \omega_S)^2 \tau_S{}^2} \tag{16}$$

$$R_2{}^{sc} = n_S \frac{4\pi^2}{3} J_{IS}{}^2 S(S + 1) \left[\tau_S + \frac{\tau_S}{1 + (\omega_I - \omega_S)^2 \tau_S{}^2} \right] \tag{17}$$

TABLE V

Incidence and Scope of Scalar Coupling Relaxation

Zeeman-separated cases where $(\omega_I - \omega_S)^2 \tau_S^2 \gg 1$

Molecule	Temperature (°C)	J_{IS} (Hz)	$T_{I(S)}$ (μs)	R_1^{sr} (s⁻¹)	R_2^{sc} (s⁻¹)	Percentage of Total Relaxation	Ref.
$^{11}B_4H_{10}$							
$B_{1,3}$	−20	—	—	—	12	44	Weiss and Grimes (1978)
$B_{2,4}$	−20	—	—	—	13	16	Weiss and Grimes (1978)
$^{119}SnCl_4$	25	470	22	—	625	100	Sharp (1972)
	25	470	22	0.63	—	100	Sharp (1972)
$^{119}SnI_4$	150	940	0.15	—	63	100	Sharp (1972)
	150	940	0.15	0.28	—	14[a]	Sharp (1972)
	220	—	—	3.3	—	100	Sharp (1972)
$^{207}PbCl_4$	25	710	7.15	—	730	100	Hawk and Sharp (1974)
	25	710	7.15	0.31	—	100	Hawk and Sharp (1974)

Molecule	Temperature (°C)	J_{IS} (Hz)	$T_{I(S)}$ (μs)	R_1^{sr} (s⁻¹)	R_1^{sc} (s⁻¹)	Percentage of Total Relaxation	Ref.
$^{13}CH_3{}^{81}Br$	5	32	1.1	—	0.0018	3	Lassigne and Wells (1974b)
	5			0.048		71	Lassigne and Wells (1974b)
$^{13}CH_3I$	24	60	0.25 (estimated)	—	0.02	—	Farrar et al. (1972)
Near-degeneracy cases where $(\omega_I - \omega_S)^2 \tau_S^2 \ll 1$							
$^{13}CH_3{}^{79}Br$	5	30	1.05	—	0.060	48	Farrar et al. (1972)
	5			0.048		38	Farrar et al. (1972)
Bromobenzene	38	—	—		0.28	—	Levy (1972)
$^{13}C^{79}Br$	125	—	—		0.83	—	Levy (1972)

[a] R_1^{sc} at 1.7 s⁻¹ is unusually large because of rapid ^{127}I relaxation.

Compared with the inconsequential 15% difference between R_1^{sa} and R_2^{sa} previously noted, scalar coupling provides an effective mechanism for transverse relaxation (R_2) that is not available for longitudinal relaxation (R_1). The subtle difference between the τ_S-containing terms in Eqs. (16) and (17) provides the basis for this distinction.

a. The Zeeman-Separated Case. Differences in Larmor frequencies for coupled nuclei represented by $\omega_I - \omega_S$ are typically 10^7–10^8 rad s^{-1}, whereas the shortest τ_S (i.e., $T_{1(S)}$ values occur for ^{35}C1, ^{79,81}Br, and ^{127}I and are typically longer than 10^{-6} s. The $[1 + (\omega_I - \omega_S)^2\tau_S^2]$ denominator in Eqs. (16) and (17) is therefore much greater than unity, and scalar coupling contributes significantly only to transverse relaxation represented by Eq. (17). The impact on transverse relaxation only is vividly portrayed in the ^{207}Pb relaxation reported by Hawk and Sharp (1974) for PbCl$_4$. At 25°C in a 1.69-T field, Eq. (16) for the ^{207}Pb relaxation reduces to

$$R_1^{sc} = 8.32[(2\pi \times 710 \text{ Hz})^2][(50 \times 10^6 \text{ s}^{-1})^{-2}][(7.15 \times 10^{-6} \text{ s})^{-1}]$$

$$= 0.009 \text{ s}^{-1}$$

and Eq. (17) reduces to

$$R_2^{sc} = 5.14[(2\pi \times 710 \text{ Hz})^2][(7.15 \times 10^{-6} \text{ s})] = 730 \text{ s}^{-1}$$

In the region where $(\omega_I - \omega_S)^2\tau_S^2 \gg 1$, R_1^{sc} is field-dependent, whereas R_2^{sc} is not. By lowering the field, R_1^{sc} is raised to the point where it makes a measurable contribution to the total longitudinal relaxation, and this field dependence allows separation of the scalar coupling component from the others. Once this is done, Eqs. (16) and (17) uniquely determine both $J_{207\text{Pb}^{35}\text{Cl}}$ and the quadrupole-dominated chlorine relaxation rate for a situation in which neither is directly measurable.

The temperature dependence of R_2^{sc} is abnormal; R_2^{sc} and thus line width increases with increasing temperature in the same way as R^{sr}. Although temperature dependence alone does not differentiate R_2^{sc} from R_2^{sr}, an independently measured T_1 whose value in the region of motional narrowing is longer than is suggested by the line width indicates additional broadening through R_2 and the presence of scalar coupling relaxation. Although the maximum strength of the scalar coupling constant at 10^4 Hz is several orders of magnitude weaker than the other coupling interactions that promote relaxation, the correlation time τ is longer by $\sim 10^{5\pm1}$ than typical τ_θ values in the range 10^{-10}–10^{-12} s.

b. The Near-Degeneracy Case. Organic bromides represent a notable exception to the Zeeman-separated case of scalar coupling relaxation. Compared with the ω differences of 10^7 to 10^8 that exist between most different nuclei, the magnetogyric ratios of ^{13}C and ^{79}Br are fortuitously similar, with $\omega_I - \omega_S = 9 \times 10^5 \text{ s}^{-1}$. Taken together with a very short ^{79}Br relaxation time of 10^{-6} s related to its quadrupole moment, the $(\omega_I - \omega_S)^2\tau_S^2$ term in Eqs. (16) and (17) is less than 1, placing scalar coupling relaxation of ^{13}C in the near-degeneracy category with the following consequences.

(a) $R_1^{sc} = R_2^{sc}$ and one obtains no additional information by measuring both relaxation rates. In particular, additional broadening resulting in the Zeeman-separated case from $R_2^{sc} > R_1^{sc}$ is not observed.

(b) R_1^{sc} is directly proportional to temperature, distinguishing this case from the Zeeman-separated one where R_1^{sc} is inversely proportional to temperature.

(c) R_1^{sc} is *not* inversely proportional to $(\omega_I - \omega_S)^2\tau_S^2$ and is therefore several orders of magnitude larger than R_1^{sc} in the Zeeman-separated case. If, however, R_1^{sc} is found to be field or frequency-dependent, then it lies in the Zeeman-separated limit.

V. Relaxation of Quadrupolar Nuclei

Nuclei with spin quantum numbers of $\frac{1}{2}$ do not couple with an electric field gradient, and as a result their positive charge is regarded as being spherically distributed. Nuclei with spin number $I > \frac{1}{2}$ all couple with an electric field gradient, and the nuclear property to which the strength of this coupling is proportional is the nuclear electric quadrupole moment Q having dimensions of (distance)^2 and measured in units of 10^{-28} m^2, referred to by early nuclear physicists as the barn unit because of its area dimensions. Where the spheroidal nuclear charge distribution is prolate or cigar-shaped, Q is assigned a positive value, and oblate or pancake-shaped distributions are assigned negative ones.

The Q values of nuclei that interest the NMR spectroscopist span four orders of magnitude from 6.9×10^{-4} barns for ^6Li to ~3 barns for ^{181}Ta and are conveniently divided into three categories in terms of size, as in Table VI.

Of the five nuclear magnetic relaxation mechanisms available, the quadrupolar mechanism is several orders of magnitude more effective than any of the others. It is this rapid relaxation that usually dominates the NMR spectroscopy of quadrupolar nuclei by making their line widths very much broader than those for $I = \frac{1}{2}$ nuclei and by drawing a relaxation curtain across most of the scalar couplings that may exist within molecules.

TABLE VI

Quadrupole Moments of Magnetically Active Nuclei Classified by Magnitude

Low-Q Nuclei, $Q < 0.1/10^{28}\ \mathrm{m}^{-2}$		Medium-Q Nuclei, $0.1/10^{28}\ \mathrm{m}^{-2} \leq Q \leq 1.0/10^{28}\ \mathrm{m}^{-2}$				High-Q Nuclei, $1.0/10^{28}\ \mathrm{m}^{-2} < Q$	
^{6}Li	6.9×10^{-4}	^{71}Ga	0.11	—		^{97}Mo	1.1
^{2}H	2.7×10^{-3}	^{131}Xe	−0.12	^{95}Mo	−0.12	^{113}In	1.14
^{133}Cs	-3×10^{-3}	^{87}Rb	0.13	—		^{115}In	1.16
		^{83}Kr	0.15	^{67}Zn	0.15		
^{14}N	0.016	^{23}Na	0.15	^{65}Cu	−0.15		
		^{27}Al	0.15	^{63}Cu	−0.16		
^{17}O	−0.026	^{69}Ga	0.18	—		^{191}Ir	1.5
		^{135}Ba	0.18	—			
^{7}Li	−0.03	—		—		^{193}Ir	1.5
^{11}B	0.036	^{73}Ge	−0.2	^{93}Nb	−0.2	^{187}Re	2.6
		^{43}Ca	0.2	^{139}La	0.21		
		^{25}Mg	0.22	^{45}Sc	−0.22		
^{9}Be	0.052	^{85}Rb	0.27	—		^{185}Re	2.8
		^{137}Ba	0.28	—			
^{39}K	0.055	^{81}Br	0.282	—		^{177}Hf	3
^{37}Cl	−0.064	^{75}As	0.3	^{51}V	0.3	^{179}Hf	3
		^{79}Br	0.332	^{99}Tc	0.3		
^{33}S	−0.064	^{87}Sr	0.36	—		^{181}Ta	3
^{41}K	0.067	^{209}Bi	−0.4	^{59}Co	0.4		
				^{201}Hg	0.50		
^{10}B	0.074	^{121}Sb	−0.4	^{55}Mn	0.55		
		^{127}I	0.6	^{197}Au	0.59		
^{35}Cl	0.080	^{123}Sb	−0.7	^{189}Os	0.8		

A. Relaxation of High-Q Nuclei

Nuclear magnetic resonance signals from high-Q nuclei are observed only for the limited group of molecules in which the nucleus is located in a highly symmetric tetrahedral (T_d) or octahedral (O_h) environment. We cannot expect to characterize a broad range of molecular structures containing these nuclei through their NMR spectra because our inability to observe the resonances is rooted in the Heisenberg uncertainty principle. At the present stage of development, no chemically relevant observation of either Hf isotope has been reported. A ^{181}Ta resonance 10,000 Hz in width and assignable to TaF_6^- in a HF–HNO$_3$ solution has been reported. (Erich *et al.*, 1973)

B. Relaxation of Medium-Q Nuclei

Medium-Q quadrupolar nuclei provide the large group whose resonances are sufficiently narrow to be observable in a representative range of chemical environments and yet are sufficiently broad that the quadrupolar mechanism can be deemed to dominate the relaxation. Metal atoms tend to occur at the centers of molecules or complexes where electric field gradients are likely to be lower than at the extremities. Nonmetals on the other hand are usually incorporated as terminal atoms, and it is worth noting that the only nonmetal isotopes among this large group of medium-Q nuclei are ^{81}Br, ^{79}Br, and ^{127}I. Considerable effort directed at their NMR spectroscopy has confirmed that in most compounds their quadrupolar relaxation rates are indeed sufficiently rapid to broaden the lines beyond detectability.

In typical chemical environments all the other members of this class yield signals that, although ranging up to several kilohertz in width, are still observable. Equation (6) indicates that the relaxation time, measured either explicitly or obtained implicitly from the line width, is dependent upon the electric field gradient at the nuclear position (stated or measured in terms of the nuclear quadrupole coupling constant) and upon the rotational correlation time. An independent measure of one of these molecular parameters provides a value for the other through a measured relaxation time, but until recently the independent measure has not been available. The double-spin probe technique (Kintzinger and Lehn, 1974) in which the relaxation rates of two different isotopes in the same molecule are measured allows one to evaluate the rotational correlation time. Less precise but more generally applicable is a calculation of the rotational correlation time using one of the theoretical models gradually being perfected (Boeré and Kidd, 1983). In either of these ways, a NQCC can be obtained from a relaxation time, and where a reliable (Kidd, 1981) Q value is available, electric field gradients at specific locations in a molecule can be obtained. Studies in which this has been done for ^{93}Nb (Tarasov *et al.*, 1978), ^{55}Mn (Ireland *et al.*, 1976), and ^{59}Co (LaRossa and Brown, 1974) have been reported.

C. Relaxation of Low-Q Nuclei

Nuclei with low values for Q give relatively narrow signals, even in molecular environments where the electric field gradient is substantial. If the field gradient is moderate or low, relaxation via the quadrupolar mechanism will be fairly slow and other mechanisms contributing to the total relaxation process must be assessed. Wehrli (1978) has provided an excellent example of this analysis in which the mechanisms contributing to the relaxation of ^{6}Li, ^{133}Cs, and ^{9}Be in both high- and low-symmetry environments have been separately identified.

By combining a good theoretical picture with consistent empirical evidence, some general conclusions about the relaxation of low-Q nuclei can be drawn.

(a) In chemical environments characterized by low electric field gradients (high symmetry), the rate of quadrupolar relaxation can be as low as 10^{-4} s^{-1}. If the atom is also small enough to accommodate the close approach of other dipolar nuclei, the rate of dipole–dipole relaxation can match or exceed this value, as happens with ^{6}Li in $Li(H_2O)_6^+$, ^{14}N in NH_4^+, ^{7}Li in $Li(H_2O)_6^+$, ^{11}B in BH_4^-, ^{9}Be in BF_4^{2-}, and ^{10}B in BH_4^- (Hertz *et al.*, 1971; Wehrli, 1978).

(b) Above temperatures of $\sim 50°C$, the rate of spin–rotation relaxation exceeds 10^{-4}–10^{-3} s^{-1} in small, symmetric molecules, and the fractions of dipolar and quadrupolar relaxation diminish as spin–rotation sustains an increasing share.

(c) In spite of their low quadrupole moments, the stereochemistries of ^{17}O, $^{35,37}Cl$, and ^{33}S are such that they seldom encounter environments of high symmetry. The quadrupolar mechanism dominates their relaxation under most circumstances and invests their relaxation rates with valuable field gradient information.

VI. The Future

The first 20 yr of NMR spectroscopy was a period characterized by a simplistic but nonetheless pragmatic view of nuclear relaxation. Quadrupolar nuclei relaxed rapidly, and narrow signals were observed only in electronic environments having high symmetry. Spin-$\frac{1}{2}$ nuclei relaxed through the inefficient dipolar mechanism, and relaxation-enhancing techniques for preventing saturation were necessary before high rf power levels could be used.

During the past 10 yr a more sophisticated understanding of relaxation has been acquired. It is now recognized that low-Q quadrupolar nuclei can give remarkably sharp lines, and great strides have been made, for example, in determining the structures of metal isopoly- and heteropolyoxyanions from highly resolved ^{17}O spectra. For spin-$\frac{1}{2}$ nuclei, the NOE test has revealed an increasing number of molecules in which spin–rotation, scalar coupling, and shielding anisotropy make significant or dominant contributions to the relaxation, thereby uncovering additional molecular constants previously hidden from view.

The next decade is likely to see a burgeoning interest in relaxation rates along with the unquestioned growth that will occur in the literature on the two traditional NMR parameters. Reliable relaxation rates are more readily extracted from Fourier transform than from continuous-wave spectra; indeed, a qualitative estimate of their magnitude is essential for the measurement of Fourier transform spectra. Increasing familiarity provides deeper insight, and in tackling both structural and dynamic problems, particularly for large molecules, relaxation rates are already providing valuable information at the margins where chemical shifts and coupling constants cease to be adequately resolved.

References

Abragam, A. (1961). "The Principles of Nuclear Magnetism," Chapter 8. Oxford University Press, London.

Aksnes, D. W., Rhodes, M., and Powles, J. G. (1968). *Molec. Phys.* **14**, 333.

Bloembergen, N. (1961). "Nuclear Magnetic Relaxation." Benjamin, New York.

Boeré, R. T., and Kidd, R. G. (1983). *In* "Annual Reports on NMR Spectroscopy" (G. A. Webb, ed.), Vol. 13, pp. 319–385. Academic Press, New York.

Debye, P. (1929). "Polar Molecules." Dover, New York.

Deverall, C. (1970). *Mol. Phys.* **18**, 319.

Edward, J. T. (1970). *J. Chem. Educ.* **47**, 261.

Einstein, A. (1917). *Phys. Z.* **18**, 121.

Erich, L. C., Gossard, A. C., and Hartless, R. L. (1973). *J. Chem. Phys.* **59**, 3911.

Farrar, T. C., Druck, S. J., Shoup, R. R., and Becker, E. D. (1972). *J. Am. Chem. Soc.* **94**, 699.

Gierer, A., and Wirtz, K. (1953). *Z. Naturforsch.* **A8**, 532.

Hawk, R. M., and Sharp, R.R. (1974). *J. Chem. Phys.* **60**, 1009.

Hertz, H. G., Tutsch, R., and Versmold, H. (1971). *Ber. Bunsenges. Phys. Chem.* **75**, 1177.

Hu, C. M., and Zwanzig, R. (1974). *J. Chem. Phys.* **60**, 4354.

Hubbard, P. S. (1963). *Phys. Rev.* **131**, 1155.

Huntress, W. T. (1970). *Adv. Magn. Reson.* **4**, 2.

Ireland, P. S., Deckert, C. A., and Brown, T. L. (1976). *J. Magn. Reson.* **23**, 485.

Jacobsen, J. P., and Schaurenberg, K. (1977). *J. Magn. Reson.* **28**, 191.

Kidd, R. G. (1981). *J. Magn. Reson.* **45**, 88.

Kintzinger, J. P., and Lehn, J. M. (1974). *J. Am. Chem. Soc.* **96**, 3313.

Kivelson, D., Kivelson, M. G., and Oppenheim, I. (1969). *J. Chem. Phys.* **52**, 1810.

LaRossa, R. A., and Brown, T. L. (1974). *J. Am. Chem. Soc.* **96**, 2072.

Lassigne, C. R., and Wells, E. J. (1977a). *J. Magn. Reson.* **27**, 215.

Lassigne, C. R., and Wells, E. J. (1977b). *Can. J. Chem.* **55**, 1303.

Levy, G. C. (1972). *J. Chem. Soc. Chem. Commun.*, p. 352.

Levy, G. C., and Cargioli, J. D. (1974). *In* "Nuclear Magnetic Resonance Spectroscopy of Nuclei Other Than Protons" (T. Axenrod and G. A. Webb, eds.), p. 251. Wiley, New York.

Levy, G. C., Holloway, C. E., Rosanske, R. C., and Hewitt, J. M. (1976). *Org. Magn. Res.* **8**, 643.

McClung, R. E. D., and Kivelson, D. (1968). *J. Chem. Phys.* **49**, 3380.

Maryott, A. A., Farrar, T. C., and Malmberg, M. S. (1971). *J. Chem. Phys.* **54**, 64.

Rugheimer, J. H., and Hubbard, P. S. (1963). *J. Chem. Phys.* **39**, 552.

Saito, H., Mantsch, H. H., and Smith, I. C. P. (1973). *J. Am. Chem. Soc.* **95**, 8453.

Sharp, R. R. (1972). *J. Chem. Phys.* **57**, 5321.

Tarasov, V. P., Privalov, V. I., and Buslaev, Yu. A. (1978). *Mol. Phys.* **35**, 1047.

Wasylishen, R. E., Pettitt, B. A., and Danchura, W. (1977). *Can. J. Chem.* **55**, 3602.

Wehrli, F. W. (1978). *J. Magn. Reson.* **30**, 193.

Weiss, R., and Grimes, R. N. (1978). *J. Am. Chem. Soc.* **100**, 1401.

Youngren, G. K., and Acrivos, A. (1975). *J. Chem. Phys.* **63**, 3846.

B

Selected Features

6 Cation Solvation

Robert G. Bryant

Chemistry Department
University of Minnesota
Minneapolis, Minnesota

I. Introduction

Solvation in general is of great importance to our understanding and sometimes misunderstanding of chemical processes in solution. The success of relatively simple thermodynamic reasoning in accounting for gas phase reactivity in ion–molecule reactions points out that much of the chemistry could be fairly well understood were it not for the solvent (Bowers, 1979). Such strategies fail in many cases to be useful for solutions, for example, in explaining the relative acidity of amines and the like (Arnett, 1973). The difficulty clearly lies in the solvent–solute interaction. Understanding solvation is extremely difficult, and although at times the solvation of ions looks like an easier case, careful inspection indicates a great number of unanswered questions.

Nuclear magnetic resonance provides a direct report of the local environment experienced by the observed nucleus in basically two ways: the resonance position or chemical shift and the resonance line widths or relaxation rates. Both chemical shift and relaxation time measurements have been exploited in gaining structural and dynamic information about the nature of ion solvation. In both cases a variety of measurements are available to the experimentalist, because often there is more than one

NMR OF NEWLY ACCESSIBLE NUCLEI, VOL. 1

solute isotope that may be studied (e.g., ^6Li or ^7Li and ^{35}Cl or ^{37}Cl) and often three or more resonances in the solvent that may be studied (e.g., ^1H, ^2H, ^{17}O, and ^{13}C). In electrolyte solutions both the cation and anion resonances may usually be investigated, so that with such an abundance of experimental data it is clear that our understanding of solutions falls far short of its potential given the quality of the structural and dynamic probes available. Although this chapter is restricted to cation solvation, the data collected even in the last few years are extensive. The present summary will focus on selected aspects of cation solvation, with examples chosen to be representative of major approaches to this area using NMR methods. No attempt has been made to be comprehensive.

II. Chemical Shift Measurements

A. Inert Systems

The simplest application of chemical shift measurements to cation solvation is in the identification of separate resonance lines for the metal ion coordination sphere and the bulk solution. The resolution of separate resonances requires that the lifetime of the solvent molecule in the ion coordination sphere exceed the reciprocal chemical shift difference. More precisely, $\tau > 1/\sqrt{2}\pi\delta$, where δ is the chemical shift difference in hertz and τ the mean residence time (Bovey, 1969). Although the shifts of the solute ion resonance may be very large in some cases, the solvent resonance shifts are small. Therefore this lifetime requirement limits the opportunity for resolving separated resonances for coordinated and uncoordinated solvent molecules. Examples are thus restricted to substitutionally inert complexes, and even then additional strategies are often useful, such as lowering the temperature drastically to slow the first coordination sphere exchange rate, making use of shift reagents to amplify the shift difference between coordinated and bulk resonances, or both. An early application of NMR in this context involved the use of ^{17}O resonances to determine the number of water molecules coordinated to transition-metal ions in aqueous solution using cobalt(II) ion as an ^{17}O shift reagent to move the resonance of the coordinated water peak away from the bulk solvent resonance for solutions of chromium(III) ion (Jackson et al., 1960). Other coordination numbers were obtained with less heroic efforts by exploiting the proton resonances of solvent molecules obtained often at very low temperatures and often in mixed solvents to minimize the freezing temperature (Fratiello, 1972). The observation of solvent resonances in mixed-solvent systems provides a direct and sometimes definitive approach to the general question of preferential solvation.

Understanding ion solvation in nonaqueous and mixed-solvent media is critical to a number of important chemical processes such as liquid extraction, chromatography, and electrolysis. High-resolution NMR of both the solvent and the metal ion can provide a direct method for defining the composition of the metal ion first solvation sphere. The results are most spectacular for inert complexes such as those of aluminum, because separate resonances may often be resolved for each solvate complex even in diamagnetic solutions. Ruben and Reuben (1976) observed a remarkable series of proton spectra for aluminum perchlorate dissolved in water–acetonitrile solutions and reported assignments for the water proton peaks which are split into 24 lines. Solutions of aluminum(III) ion containing water are complicated by the acid–base properties of the metal ion-coordinated water. Thus hydroxoaluminum(III) complexes had to be included in the analysis. In addition, participation of the perchlorate ion in the metal first coordination sphere was required and has since been verified by other laboratories (Akitt and Farthing, 1978; Akitt and Mann, 1981; Akitt et al., 1979). Representative water proton spectra taken at different water/aluminum(III) ion molar ratios in acetonitrile are shown in Fig. 1, and the assignments are reproduced in Table I. The use of solvent proton resonances appears to remain a valuable tool, especially when magnetic field strengths five times as large as those used for Fig. 1 are now commercially available, which provide five times the chemical shift resolution in such samples.

Although the resolution in the proton spectrum in these solutions is impressive, the chemical shifts associated with heavier nuclides are generally larger. For metals like platinum and lead, the anisotropy in the chemical shift may also lead to very severe line broadening mechanisms at high field that actually decrease resolution rather than increase it with increasing field strength (Hays et al., 1981). For aluminum and many other metal ions, including alkali and alkaline earth ions, the NMR line widths are dominated by the nuclear electric quadrupole relaxation mechanism rather than dipole–dipole or chemical shift mechanisms (Harris and Mann, 1978). If the nuclear electric quadrupole moment is large, as in ^{137}Ba, for example, the efficient relaxation broadens the lines sufficiently to provide a real limit on the potential metal ion resonance utility. In the case of many metal resonances, however, the resolution may remain quite good for low-molecular-weight complexes (Harris and Mann, 1978), and ^{27}Al provides an excellent example.

The ^{27}Al-NMR spectrum in water–acetonitrile solutions has received considerable attention. Akitt and co-workers (1979) have reported on the first coordination sphere composition, hydrolysis of the metal–solvent complex, and participation of anionic ligands such as chloride and perchlorate ions in the aluminum(III) ion first coordination sphere. The hy-

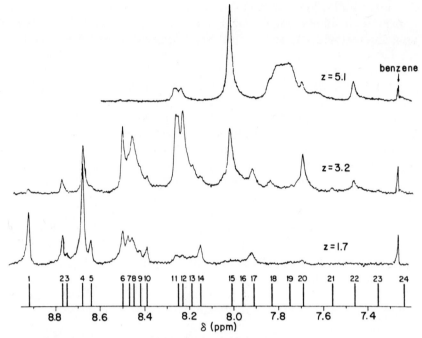

Fig. 1 The water proton region of the ^1H NMR spectrum of aluminum perchlorate dissolved in water–acetonitrile solutions at 258 K as a function of Z, the water/aluminum ion ratio. Peak assignments are indicated in Table I. Reprinted with permission from Ruben and Rueben (1976), copyright 1976, American Chemical Society.

drolysis problem is more complex than it first appears. The product is apparently a function of the sample history or the chemical reaction that generates the deprotonated product ion species (Akitt and Farthing, 1978; Akitt and Mann, 1981). The product of preparation with sodium carbonate is the complex species formulated as $AlO_4Al_{12}(OH)_{24}(H_2O)_{12}^{7+}$ but, depending on the history of the sample, a distribution of other structures is indicated by ^{27}Al NMR among other spectroscopic approaches.

Recently, Wehrli and Wehrli (1981) reported ^{27}Al spectra obtained at high field that demonstrated beautifully the potential of the metal ion resonance. The spectrum, with their assignments, obtained from aluminum(III) ion in acetonitrile solutions is shown in Fig. 2.

The resolution of chemically distinct environments for the solvent, coordinated and uncoordinated, provides a direct measure of the coordination number of the ion through the integrated intensity of the signal. However, in principle if not often in practice, the geometry of the complex with the solvent complex is ambiguous and is assigned by general

TABLE I

Peak Assignment in the Spectral Region of Water Coordinated to Aluminum(III) at −15°C

Peak	Shift (ppm)	Complex[a] x	i	p	Structure[b]
1	8.92	1	5	—	—
2	8.77	2	4	—	W,W-trans
3	8.75	1	4	1	W,P-trans
4	8.68	2	4	—	W,W-cis
5	8.64	1	4	1	W,P-cis
6	8.50	3	3	—	W,W,W-cis,trans
7	8.47	2	3	—	W,W,-trans
8	8.45	3	3	—	W,W,W-cis,cis
9	8.42	2	3	1	W,W-cis, P-cis,cis
10	8.39	2	3	1	W,W-cis, P-cis,trans
11	8.25	4	2	—	AN,AN-cis
		3	2	1	AN,AN-trans
12	8.23	4	2	—	AN,AN-trans
13	8.19	3	2	1	AN,AN-cis, P-cis,trans
14	8.15	3	2	1	AN,AN-cis, P-cis,cis[c]
15	8.01	5	1	—	In-plane water
16	7.96	4	1	1	AN,P-trans
17	7.91	4	1	1	AN,P-cis
18	7.83	5	1	—	Apical water
19	7.75	6	—	—	—
20	7.69	5	—	1	—
21	7.56 Monohydroxo complexes				—
22	7.46 Monohydroxo complexes				—
23	7.35 Dihydroxo complexes				—
24	7.25 Dihydroxo complexes				—

[a] $Al(H_2O)_x(CH_3CN)_i-(ClO_4)_p^{3-p+}$, $x = 6 - i - p$.
[b] W, H_2O; AN, acetonitrile; P, perchlorate.
[c] This peak may also contain contribution from an $Al(H_2O)_2(CH_3CN)_2(ClO_4)^{2+}$ complex with a bidentate perchlorate.

reference to compounds for which the structure is accurately known. It is indeed rare that one cannot infer the solvate structure from the coordination number, but a more direct conclusion may be drawn when the scalar coupling between a solvent nucleus and the metal atom is observed. Similar dynamic constraints are required with respect to solvent molecule lifetime in the metal coordination sphere of the metal ion, but the scalar coupling provides definitive evidence about both the number and geometry of the solvent complex when such fortunate circumstances arise. Fig-

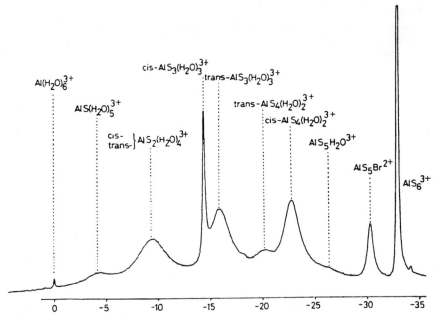

Fig. 2 The 93.8-MHz ^{27}Al-NMR spectrum of 1 M aluminum bromide in acetonitrile following the addition of 20 μl of water to make the molar ratio of water to total aluminum ion 0.5. The identification of the mixed solvate peaks from $[AlS_n(H_2O)_{6-n}]^{3+}$, where S is CH_3CN, is discussed by Wehrli and Wehrli (1981). Reprinted with permission, copyright 1981, Academic Press.

ure 3 shows a clear example of the ^{27}Al-NMR spectrum in nitromethane solutions where the two different phosphoramides were used, both of which can also be used as the solvent (Delpuech *et al.*, 1975). In this case the coupling between the phosphorus and the aluminum leaves no doubt, from the resolution of the five-line pattern in the case of hexamethylphosphorotriamide (HMPA) and the seven-line pattern in the case of trimethyl phosphate (TMPA), that the aluminum first coordination sphere includes four and six equivalent phosphorus atoms, respectively. In addition, the temperature dependence of the multiplet provides an accurate and straightforward means of extracting the solvent lifetimes and activation parameters for first coordination sphere solvent exchange on the metal.

B. Labile Systems

Substitutionally inert metal ion complexes are relatively rare. Usually the solvent lifetime in the metal ion first coordination sphere is short

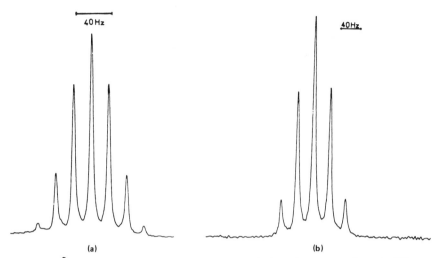

Fig. 3 The ^{27}Al-NMR spectra of a nitromethane solution of Al(TMPA)$_6^{3+}$ and 3ClO$_4^-$ at 298 K (a) and Al(HMPA)$_4^{3+}$ and 3ClO$_4^-$ at 243 K (b) recorded at 22.6 MHz by Delpuech *et al.* (1975). Reprinted with permission from Delpuech *et al.* (1975), copyright 1975, American Chemical Society.

compared with the reciprocal of the chemical shift experienced by a solvent nucleus on coordination with the bulk solvent, so that only an exchange averaged solvent resonance is observed. This resonance may still yield very useful information. Of course, one of the earliest applications of NMR to the ion solvation problem was exploitation of the chemical exchange broadening of solvent resonances associated with coordination to a paramagnetic transition metal to obtain the solvent molecule lifetime in the first coordination sphere (Swift and Connick, 1962). Applications including various solvent resonances are numerous and are now more often discussed in the context of metal ion substitution mechanisms. Some interesting recent applications of this technique have included the pressure dependence of the solvent exchange reaction (Sisley *et al.*, 1982; Meyer *et al.*, 1982). A more complete understanding of the solvation of all the participants in a chemical reaction would make interpretation of the activation parameters easier. In particular, a more detailed understanding of the second coordination sphere solvation of a transition-metal complex with no solvent in the first coordination sphere would be extremely helpful (Langford and Tong, 1977). In cases where the exchange of the solvent with the ions involved is rapid compared with the chemical shifts on association, interpretation of the NMR measurements becomes model-dependent. To make progress one has to deal in one way or another with the fact that both cations and anions may influence the chemical shift of the solvent resonances, and that cations and anions form ion pairs in many solvents so that the separation of solvation phenomena from ion association phenomena becomes important. In spite of the difficulty, such measurements have received a great deal of attention.

In electrolyte solutions of a single solvent, the concentration dependence of the chemical shift extrapolated to infinite dilution permits comparison of one solvent with another. Vogrin and Malinowski (1975) provide an example of a strategy for extracting solvation numbers from chemical shifts of the proton resonance of the solvent. In a somewhat different approach Miura and Fukui (1980) incorporate calculations of proton shieldings induced by ions in acetone solutions and conclude that there is no anion–water association in acetone.

Direct observation of one of the solute resonances eliminates in part the difficulty in dividing up the contributions from the anion and the cation to the averaged solvent shifts. By extrapolating chemical shifts of solute resonances to infinite dilution, one can obtain ion pair-independent shifts for a solvated ion surrounded by a solvent complex of one solvent molecule type. Thus Popov and co-workers, for example, have measured the chemical shifts of a variety of cation resonances as a function of salt concentration in many solvents (Farmer and Popov, 1981; Heubel and Popov, 1979; Li and Popov, 1982; Popov, 1978, 1979; Shih and Popov, 1977). The infinite dilution shifts correlate well with measures of a solvent molecule's ability to donate electrons to a cation, such as the Gutmann donor number (Gutmann and Wychera, 1966; Gutmann, 1968). Higher concentrations of the solute usually induce very significant shifts in the cation resonances, which are anion-dependent. It is interesting to note that these effects, which are attributed to ion pairing phenomena, do not become insignificant at concentrations where the ion concentration is sufficiently dilute for the solution to be treated by limiting theoretical approaches.

In many instances there is concern with the issue of preferential solvation, that is, whether the composition of the solute ion first coordination sphere is identical to the composition in the bulk of a mixed-solvent system or whether one solvent component is preferentially bound to the ion. In transition-metal systems where the bonding is not very dominantly electrostatic, one anticipates that preferential solvation is the rule and finds the great preference of cobalt systems for nitrogen rather than oxygen, for example (Silen and Martell, 1964). However, with ions that we think are simpler, such as alkali metal ions, preferential interactions may not be taken for granted. Analysis of a chemical shift that represents an average over a number of species must make approximations about the behavior of the chemical shift as a function of the solvation sphere composition. The simplest approximation is to assume that the chemical shift changes linearly with the composition of the cation first solvation sphere as one type of solvent molecule is changed into another (Frankel *et al.*, 1965, 1970). There are several reasons why one might expect this assumption to fail, such as different chemical shifts arising from nonlinear changes in the electronic response to subtle symmetry changes resulting from a mixed-ligand solvation of the cation. In some instances the approximation does fail (Henrich *et al.*, 1977); however, it has been remarkably useful in the analysis of solute resonance shifts in mixed-solvent solutions. The treatment of preferential solvation may be visualized as an association competition at the cation site in which the different solvent species are each characterized by stepwise association constants. The preferential solvation problem is thus the same as a ligand displacement problem and involves all of the same difficulties. Covington and co-workers have developed a data treatment scheme that includes the multiple equilibrium in a careful and systematic way. Many have exploited this analysis in the treatment of solute cation resonances in mixed-solvent systems (Briggs and Hinton, 1979; Covington and Covington, 1975; Covington and Newman, 1979; Covington and Thain, 1974; Covington *et al.*, 1973a, b, 1974; Greenberg and Popov, 1975; Laszlo, 1978). In the best cases it is possible to extract the constants that characterize the preferential solvation along with an assessment of the usual assumption that chemical shift is a linear function of the solvent composition in the first coordination sphere (Covington and Newman, 1976).

Recently Laszlo and co-workers have adopted a slightly different formalism for treating such magnetic resonance data according to methods originally developed by Hill (1910) and extensively applied to multiple-ligand equilibria in biochemistry such as the binding of oxygen to hemoglobin (Gill *et al.*, 1978). This data analysis strategy (Delville *et al.*, 1980) requires the assumption that the chemical shift of the solute resonance is a linear function of the composition of the solvation sphere and that the concentration of free competing solvent molecules is known, which is to a good approximation equal to the total concentration of the potential ligands for solutions dilute in metal ion. For a maximum coordination number of 4, the fraction of metal sites Y occupied by one of the solvent molecules of interest is given by

$$Y = (\delta_{obs} - \delta_0)/(\delta_4 - \delta_0) \qquad (1)$$

A plot in $\ln [Y/(1 - Y)]$ versus $\ln X$, where X is the mole ratio of the competing ligand concentrations, is then a straight line if all the intrinsic association constants are equal, but it curves in the case where there is a shift or cooperativity for the addition of more than one ligand. The data for a series of binary solutions of monoamines in tetrahydrofuran (THF)

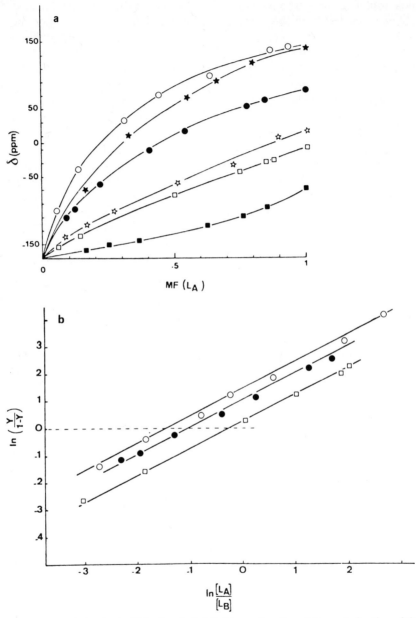

Fig. 4 The variation in the ^{23}Na chemical shift (a) as a function of the mole fraction of the amine for binary solvent mixtures of aniline (■), pyridine (□), piperidine (★), pyrrolidine (●), propylamine (○), and isopropylamine (☆) in THF. The Hill plots for three of these solvent systems are shown in (b). (Delville *et al.*, 1981a). Reprinted with permission.

are shown in Fig. 4 along with representative Hill plots (Delville *et al.*, 1981a). These data were analyzed to give the intrinsic equilibrium constants K and the limiting shifts so that the composition of the solvation sphere could be calculated as a function of the bulk solvent composition.

The particular advantage of this strategy is in the case where the solvent is bifunctional, e.g., a diamine. The Hill formalism then produces a curved plot as in Fig. 5, which may be fit to extract the effective intrinsic stepwise equilibrium constants (Delville *et al.*, 1981b). These authors argue that this method provides an excellent measure of the chelate effect that does not depend on difficult choices of appropriate controls and standard states. Further applications of this approach will be important in considering the nuclear relaxation of systems dominated by the nuclear electric quadrupole moment.

An important aspect of cation solvation is the nature of the solvation that remains when the metal ion is partially complexed by other nonsolvent ligands. A particularly interesting case is the complexation and solvation of ions in macrocyclic systems where the chemical shift of the metal resonance as a function of solvent may indicate whether the solvent has access to the metal or not. Popov and Lehn (1979) and co-workers (Kauffmann *et al.*, 1980) have studied such systems extensively and distinguish two classes of macrocyclic complexes by studying the cation resonances: those that exclude solvent completely and those that do not. The former case is called inclusive, the latter exclusive. The criterion for distinguishing between the two classes is the dependence or indepen-

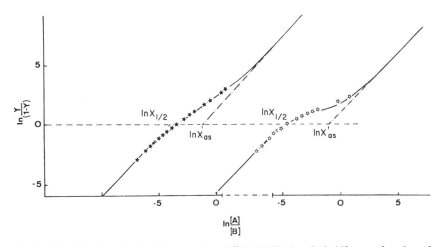

Fig. 5 The Hill plots obtained from studies of ^{23}Na-NMR chemical shifts as a function of the mole fraction of ethylenediamine (left) and diethylenetriamine (right) in trihydrofuran (Delville *et al.*, 1981b). Reprinted with permission.

dence of the cation resonance on the solvent. This type of study also lends support to the idea that the chemical shift is strongly dominated by the nearest-neighbor atoms.

III. Nuclear Magnetic Relaxation

Nuclear magnetic relaxation provides a direct and powerful experimental approach to the study of cation solvation (Bryant, 1978). The recovery of nuclear magnetization to equilibrium following a perturbation depends on where the perturbation leaves the net magnetization and in what direction the decay or recovery is observed. The recovery of the magnetization parallel to the applied field direction is usually called longitudinal relaxation and is characterized by the time constant T_1. The decay of magnetization in a direction perpendicular to the applied field direction is usually called transverse relaxation and is characterized by the time constant T_2. Other measurements are possible, but this discussion will focus on these two. Very crudely, the longitudinal relaxation rate is a measure of the efficiency of energy transfer between the nuclear spin system and other degrees of freedom in the sample and depends on motions at approximately the Larmor precession frequency or faster. The transverse relaxation rate is generally equal to or greater than the longitudinal relaxation rate and may be affected by processes that are very much slower than the Larmor precession such as a chemical exchange event that changes the magnetic environment of the observed nucleus. In the case of low-molecular-weight solutes, the relaxation discussion is simplified because the dominant motions in the solution are all rapid compared with the Larmor precession frequency, so that a condition of extreme narrowing of the resonance lines obtains (Abragam, 1961). This represents a considerable simplification of the theoretical apparatus required to interpret data, particularly for metal resonances of nuclei with spin greater than 1 (Bull *et al.*, 1979). Although there are a number of nuclear magnetic relaxation mechanisms, two have dominated applications to the study of solutions: the dipole–dipole mechanism and the nuclear electric quadrupole mechanism.

A. Dipole–Dipole Relaxation

Although the transverse relaxation is often directly related to the NMR line width by the relation $\Delta\nu = 1/\pi T_2$, most applications focus on the more accurately measured longitudinal relaxation rates. For the dipole–dipole interaction the longitudinal relaxation rate simplifies to

$$1/T_1 = 2\gamma^4\hbar^2 I(I + 1)\tau_c/r^6 \tag{2}$$

where I is the nuclear spin, γ the magnetogyric ratio, and τ_c the reorientation of the intermoment vector r. When unlike spins are involved, additional spectral densities are required. Although nonexponential relaxation is predicted, requiring solution and analysis of coupled equations (Solomon, 1955), the situation often simplifies to an exponential case for the observed spin I in the presence of a second spin S.

The relaxation rate contains the intermoment distance and the time constant characterizing its reorientation. Often the importance of both inter- and intramolecular contributions makes a unique interpretation of the relaxation rate difficult; however, a study of the relaxation as a function of the isotope composition of the solution, substituting deuterium for hydrogen, for example, may permit separation of the intermolecular contribution which by design may characterize a solvent–solute interaction. Alternatively, measurements of the deuterium relaxation rate may give a measure of intramolecular relaxation rates directly, because the deuterium is relaxed by the nuclear electric quadrupole interaction, which is an entirely intramolecular mechanism. Once the intermolecular contribution to the solvent or solute relaxation rate is isolated, it may be analyzed in various ways to characterize the structure and dynamics of the ion solvation. This approach has been exploited and developed by Hertz and co-workers at Karlsruhe who analyze the intermolecular relaxation in terms of the nature of the pair distribution functions for the interaction particles. The intermolecular contribution to the relaxation may be written

$$1/T_1 = \tfrac{4}{3}\gamma_I^2\gamma_S^2\hbar^2 S(S + 1)\tau_c^* \int_a^\infty 4\pi r_0^2[P(r_0)/r_0^6] \, dr_0 \tag{3}$$

where a is the closest distance of approach between the proton and the ion and $P(r_0)$ the real proton–ion pair distribution function divided by the number density of the protons (Langer and Hertz, 1977). A model is required for $P(r)$, and a choice is indicated:

$$P(r_0) = \begin{cases} 0 & \text{for } r_0 < a \\[2mm] \dfrac{n_h P_0 (r_0 - a)^m}{1 + (r_0 - a)^m} \exp[-q(r_0 - a)] & \text{for } a \leq r \leq b \\[2mm] C_I' & \text{for } r_0 > b \end{cases} \tag{4}$$

where b is the closest distance of approach for protons not in the ion first coordination sphere, n_h the coordination number of the ion, m and q the assumed potential for the interaction, P_0 the normalization factor such that the integral of $P(r_0)$ from a to b equals the coordination number, and C'_I the number density of the ions in the solution. The procedure yields expressions for both the first coordination sphere of the ion and the sum of outer coordination sphere effects. The first coordination sphere equation becomes

$$1/T_{1,1st} = \tfrac{4}{3}\gamma_I^2\gamma_S^2\hbar^2 S(S\,+\,1)n_h\tau_c^* a^6\langle 1/r_0^6\rangle \tag{5}$$

where $\langle 1/r_0^6\rangle$ is obtained from the assumed model distribution function, $\tau_c^* \simeq a^2/3\bar{D}$, where $\bar{D}$ is the arithmetic mean of the ion and the water self-diffusion coefficient and a_d is estimated from models for the closest distance of approach for the ion and the proton. The outer-sphere contribution is then

$$1/T_{1,2nd} = \tfrac{16}{9}\,\pi\gamma_I^2\gamma_S^2\hbar^2 S(S\,+\,1)\tau_{cI}^*(1/b^3)C'_I \tag{6}$$

With such a model it is possible to analyze the intermolecular relaxation data for values of a or b, compare them with known molecular dimensions, and construct solution structures around the ion. An illustrative example is provided by the data of Geiger and Hertz (1976a, b) who investigated the water molecule orientation around the lithium ion in aqueous solutions. Based on the intermolecular proton contribution to the ^{6}Li and ^{7}Li relaxation rates, they concluded that the water dipole is not radially oriented around the lithium ion.

A more straightforward case occurs when the solute ion is paramagnetic, so that the electron nuclear dipole–dipole interaction dominates other contributions to the solvent relaxation. Both structural and dynamic information may be obtained. Paramagnetic metal ion effects on the solvent relaxation have been exploited extensively in the study of macromolecules (Burton et al., 1979). Goldammer and Kreysch (1978) have reported an interesting study on the proton relaxation rates in dilute solutions of manganese perchlorate in dimethyl sulfoxide measured over a wide frequency range. The frequency dependence of the relaxation is a powerful means for determining the correlation times for the different spectral densities in the relaxation equation, including the electron relaxation time. From this extensive data set these authors have deduced the solvation number, the intermoment distances in the solvent complex, and the correlation times for the motion of the solvent relative to the manganese ion. In a related study Delpuech and co-workers (Boubel et al., 1977) measured the proton and carbon relaxation times for the same solvent in manganese ion solutions.

B. Nuclear Electric Quadrupole Relaxation

Very often the solute resonance is dominated by the nuclear electric quadrupole moment eQ interacting with fluctuating electric field gradients eq at the observed nucleus. In the limit of extreme narrowing the longitudinal relaxation rate is given by

$$\frac{1}{T_1} = \frac{3\pi^2}{10} \frac{(2I + 3)}{I^2(2I - 1)} \frac{(e^2qQ)^2}{h} \left(\frac{1 + \eta^2}{3}\right)\tau_c \tag{7}$$

where η is the asymmetry parameter and the correlation describes the fluctuation rate of the field gradient eq. The relaxation rates of the quadrupole relaxed solute nuclei thus indicate the local electron symmetry of the ion, and major distortions such as substitution of one molecule for another in the first coordination sphere lead to large changes in the observed relaxation rate. Detailed and quantitative analysis of the relaxation rate, however, is difficult because it is difficult to determine how the electric fields and the field gradients important in the relaxation process propagate through the electron cloud surrounding the observed nucleus. The essential question is, What is the ultimate source and nature of the field gradients that drive relaxation? A good resolution of this question would be helpful in the study of mixed solvation and ion–ion association. There are two basic, rather different approaches to this problem. One is a relaxation model that treats the field gradients at the relaxing ion as arising from fluctuations in the electric environment of the ion produced by solvent dipoles and counterions, i.e., an electrostatic origin (Geiger and Hertz, 1976a, b; Hertz et al., 1969, 1974; Hertz, 1961, 1973; Valiev, 1960, 1962, 1964; Valiev and Khabibullin, 1961; Valiev and Zaripov, 1966). The other is a model that treats the field gradients as arising from short-range interactions resulting when an ion collides with a solvent or other solute particle and suffers a distortion in the symmetry of the electron cloud (Deverell, 1969a, b). This mechanism then is not electrostatic in that the field gradient is propagated by a transient rearrangement of the observed ion–electron distribution. An excellent summary of the different approaches has been provided by Lindman and Forsén (1976).

Since the formation of ion pairs in solution clearly affects both mechanisms, an often used strategy is to study the concentration dependence of the relaxation times and extrapolate to the limit of infinite dilution. The infinite dilution values may then be compared with theory. For the case where the solvent dipole, water in most cases, may reorient randomly and where cross-correlation effects are neglected, the relaxation rate at the solute ion arising from the dipole motion is given by Hertz as

$$1/T_1 = 1/T_2 = \{24\pi^3(2I + 3)(eQ/h)^2[\mu(1 + \gamma_\infty)P]^2(C_{H_2O}\tau_{H_2O})\}$$
$$\times [r_0^5 5I^2(2I - 1)]^{-1} \tag{8}$$

where μ is the water molecule dipole moment, $1 + \gamma_\infty$ the Sternheimer antishielding factor, P the polarization factor approximated by the expression $(2\varepsilon + 3)/5\varepsilon$ (where ε is the dielectric constant of the solvent), C_{H_2O} the concentration of water dipoles, τ_{H_2O} the effective correlation time for rotation of the water dipole vector, and r_0 the distance between the observed nucleus and the first coordination sphere water molecule considered a point dipole. Application of this model requires estimation of the antishielding parameters and the polarization factor, which may sometimes be difficult. Nevertheless, the model has dominated interpretation of the relaxation of solute resonances. Only a few examples will be cited.

This approach to quadrupole moment-dominated NMR relaxation as exemplified in a recent study by Weingärtner (1980) in which the ^{7}Li, ^{1}H, and ^{2}H relaxation rates over a wide temperature range, including the supercooled regime of a 6 m lithium iodide solution, are reported. Similar measurements have been reported for glycerol solutions (Geiger and Hertz, 1976a). Isotope substitutions permit isolation of the quadrupole contribution to the ^{7}Li relaxation rate which can then be compared with the proton and deuteron relaxation of the solvent. The important feature of the data is that all the relaxation rates pass through a maximum at the same temperature. This result is taken as an indication that a common reorientational Brownian process drives the relaxation for all three resonances, which suggests a molecular complex reorientation. This is consistent with the formulation of the electrostatic theory, and although the concentration of salt is high and ion pairing problems are not negligible in such systems, an explanation of quadrupole relaxation must be able to account for the fact that the solvent and the solute resonances apparently respond to the same dynamic process.

It must be pointed out that electrostatic theories, while commonly employed, are not proven in the sense that there is no reason to doubt their universal validity. On the contrary, Mishustin and Kessler (1975) and Kessler *et al.* (1977) have raised questions about small ions, such as lithium, which are assumed to be more tightly solvated and thus have higher local symmetry. In addition, Friedman (1978) has suggested a different view of the water dynamics next to the ion, which involves a wagging motion between lone-pair positions of the coordinated water which are driven in part by the hydrogen bond requirements of the second coordination sphere. A different although related argument has been advanced by Takahashi (1977) for ^{27}Al relaxation rates.

The major difficulty with respect to the suggestions of Deverell that the relaxation is dominated by electronic effects is the difficulty in making an accurate test of the theory. The theory gives the relaxation rate as

$$\frac{1}{T_1} = \frac{3\pi^2}{1000} \frac{2I+3}{I^2(2I-1)} \left(\frac{e^2Q}{h}\right)^2 \left(\frac{\sigma_p \, \Delta E}{\alpha^2}\right)^2 \tau_c \qquad (9)$$

where σ_p is the paramagnetic contribution to the chemical shielding or shift, ΔE the mean electron excitation energy, and $\alpha = e^2/hc$. The correlation time is taken as the reorientational correlation time for the solvent. The difficulty in testing this model has been overcome by Delville and coworkers (1981c) who rearrange this equation in terms of the viscosity-normalized line width $\Delta\nu^*_{1/2}$,

$$(\Delta\nu_{1/2})_i^{1/2} = C_i \overline{\Delta E_i} (\tau_c/\eta)^{1/2} (\sigma_i - \sigma_d) \qquad (10)$$

where σ_d is the diamagnetic part of the shift. The critical feature is the proportionality of the square root of the line width or the quadrupole coupling constant to the chemical shift. By studying the ^{23}Na line widths and shifts as a function of the solvent composition in binary solvent mixtures, and analyzing the data in terms of successive displacement of one solvent molecule by the other competitor, each mixed-solvent coordination complex in the stepwise displacement may be characterized by a shift and a line width. The Deverell hypothesis is then tested by plotting the square root of the line width versus the shift for each solvent molecule displaced for a variety of solvents. A representative plot is shown in Fig. 6. The intercept provides the diamagnetic part of the shielding, and the linearity of the plots is taken as a strong indication that the sodium ion relaxation rate is dominated by distortions in the electronic structure of the sodium ion rather than the propagation of electrostatic fields through the electrons to the nucleus. It is interesting to note that Kintzinger and Lehn (1974) have reported data for sodium–cryptate complexes that support the conclusions of Delville et al. (1981c).

In a very different sort of experiment, Rose and Bryant (1979) measured the effect of a charge placed external to the primary coordination sphere of a substitutionally inert cobalt(III) complex. After correction for volume effects this group also found that the cobalt relaxation did not respond to a trianion charge placed in contact with the second coordination sphere, which one would think it would if electrostatic fields dominated the cobalt quadrupole relaxation.

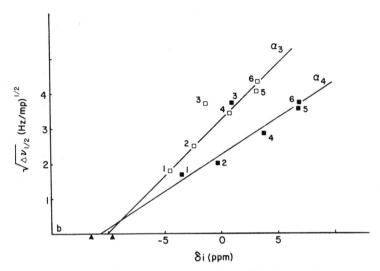

Fig. 6 The square root of the viscosity-normalized ^{23}Na line width versus the ^{23}Na chemical shifts as a function of the composition of the sodium ion solvation sphere for a series of amine solvents in THF. These plots correspond to the complexes with three amine and four amine ligands as the amine is varied. 1, Aniline; 2, pyridine; 3, piperidine; 4, pyrrolidine; 5, n-propylamine; 6, isopropylamine (Delville *et al.*, 1981c). Reprinted with permission, copyright 1981, Academic Press.

Regardless of whether we understand the fine details of the quadrupole relaxation in electrolyte solutions, the line widths and relaxation times may be exploited in studies on preferential solvation. Holz (1978), for example, reports a novel method for exploiting changes in the relaxation rates on substitution of D_2O for H_2O in alkali metal halide solutions in mixed solvents containing water with methanol, formamide, and dimethylsulfoxide. He suggests that the isotope effects on the relaxation rate may be used to study preferential solvation of the alkali cations and indicates, for example, preferential hydration of sodium and rubidium ion by water in these solvents.

It should be clear that there are several features of nuclear relaxation dominated by the nuclear electric quadrupole interaction that are unsettled. It may also be that in a sense some aspects of the argument, as it were, are semantic. That is, there is a coupling of the electronic cloud of an ion to the instantaneous motions of surrounding ions and dipoles, otherwise we would not have van der Waals complexes and the like. The question is one of degree. In summary, NMR is a powerful means for addressing both structural and dynamic aspects of ion solvation. There is no doubt that in some ways our understanding of the spectroscopy itself requires refinement in order to advance solvation studies further. Because the frequencies used are low, magnetic resonance experiments report an averaged picture—not a snapshot. Nevertheless it is clear that magnetic resonance is uniquely suited for solvation studies.

·   **References**

Abragam, A. (1961). "The Principles of Nuclear Magnetism," Chapter 8. Clarendon, Oxford.
Akitt, J. W., and Farthing, A. (1978). *J. Magn. Reson.* **32,** 345–352.
Akitt, J. W., and Mann, B. E. (1981). *J. Magn. Reson.* **44,** 584–589.
Akitt, J. W., Duncan, R. H., Beattie, I. R., and Jones, P. J. (1979). *J. Magn. Reson.* **34,** 435–437.
Arnett, E. M. (1973). *Acc. Chem. Res.* **6,** 404–409.
Boubel, J. C., Brondeau, J., and Delpuech, J. J. (1977). *Adv. Mol. Relaxation Interact. Processes* **11,** 323–336.
Bovey, F. A. (1969). "Nuclear Magnetic Resonance Spectroscopy," p. 188. Academic Press, New York.
Bowers, M. T., ed. (1979). "Gas Phase Ion Chemistry." Academic Press, New York.
Briggs, R. M., and Hinton, J. F. (1979). *J. Solution Chem.* **8,** 519–527.
Bryant, R. G. (1978). *Ann. Rev. Phys. Chem.* **29,** 167–188.
Bull, T., Forsén, S., and Turner, D. (1979). *J. Chem. Phys.* **70,** 3106–3111.
Burton, D. R., Forsén, S., Karlstrom, G., and Dwek, R. A. (1979). *Prog. NMR Spectrosc.* **13,** 1–45.
Covington, A. D., and Covington, A. K. (1975). *J. Chem. Soc. Faraday Trans. 1* **71,** 831–840.
Covington, A. K., and Newman, K. E. (1976). *In* "Advances in Chemistry Series" (W. F. Furter, ed.), Vol. 155, pp. 153–196. American Chemical Society, Washington, D.C.
Covington, A. K., and Newman, K. E. (1979). *Pure Appl. Chem.* **51,** 2041–2058.
Covington, A. K., and Thain, J. M. (1974). *J. Chem. Soc. Faraday Trans. 1* **70,** 1879–1887.
Covington, A. K., Lilley, T. H., Newman, K. E., and Porthouse, G. A. (1973a). *J. Chem. Soc. Faraday Trans. 1* **69,** 963–972.
Covington, A. K., Newman, K. E., and Lilley, T. H. (1973b). *J. Chem. Soc. Faraday Trans. 1* **69,** 973–983.
Covington, A. K., Lanthke, I. R., and Thain, J. M. (1974). *J. Chem. Soc. Faraday Trans. 1* **70,** 1869–1878.
Delpuech, J. J., Khaddar, M. R., Peguy, A. A., and Rubini, P. R. (1975). *J. Am. Chem. Soc.* **97,** 3373–3379.
Delville, A., Detellier, C., Gerstmans, A., and Laszlo, P. (1980). *J. Am. Chem. Soc.* **102,** 6558–6559.
Delville, A., Detellier, C., Gerstmans, A., and Laszlo, P. (1981a). *Helv. Chim. Acta* **64,** 547–555.
Delville, A., Detellier, C., Gerstmans, A., and Laszlo, P. (1981b). *Helv. Chim. Acta* **64,** 556–567.
Delville, A., Detellier, C., Gerstmans, A., and Laszlo, P. (1981c). *J. Magn. Reson.* **42,** 14–24.
Deverell, C. (1969a). *Prog. NMR Spectrosc.* **4,** 235–334.

Deverell, C. (1969b). *Mol. Phys.* **16**, 491–500.

Farmer, R. M., and Popov, A. I. (1981). *Inorg. Nucl. Chem. Lett.* **17**, 51–56.

Frankel, L. S., Stengle, T. R., and Langford, C. H. (1965). *Chem. Commun.* 393–394.

Frankel, L. S., Langford, C. H., and Stengle, T. R. (1970). *J. Phys. Chem.* **74**, 1376–1381.

Fratiello, A. (1972). *Prog. Inorg. Chem.* **17**, 57–92.

Friedman, H. L. (1978). *In* "Protons and Ions Involved in Fast Dynamic Phenomena" (P. Laszlo, ed.), pp. 27–42. Elsevier, Amsterdam.

Geiger, A., and Hertz, H. G. (1976a). *Adv. Mol. Relaxation Process* **9**, 293–326.

Geiger, A., and Hertz, H. G. (1976b). *J. Solution. Chem.* **5**, 365–388.

Gill, S. J., Gaud, H. T., Wyman, J., and Barisas, B. G. (1978). *Biophys. Chem.* **8**, 53–59.

Goldammer, E., and Kreysch, W. (1978). *Ber. Bunsenges. Phys. Chem.* **82**, 463–468.

Greenberg, M. S., and Popov, A. I. (1975). *Spectrochim. Acta* **31A**, 697–705.

Gutmann, V. (1968). "Coordination Chemistry in Nonaqueous Solvents." Springer-Verlag, Berlin.

Gutmann, V., and Wychera, E. (1966). *Inorg. Nucl. Chem. Lett.* **2**, 257–260.

Harris, R. K., and Mann, B. E., eds. (1978). "NMR and the Periodic Table." Academic Press, New York.

Hayes, G. R., Gillies, D. G., Blaauw, L. P., and Clague, A. D. H. (1981). *J. Magn. Reson.* **45**, 102–107.

Henrich, P. M., Ackerman, J. J. H., and Maciel, G. E. (1977). *J. Am. Chem. Soc.* **99**, 2544–2548.

Hertz, H. G. (1961). *Z. Elektrochem.* **65**, 20–36.

Hertz, H. G. (1973). *Ber. Bunsenges. Phys. Chem.* **77**, 531–540, 688–697.

Hertz, H. G., Stalidis, G., and Versmold, H. (1969). *J. Chim. Phys. Physicochim. Biol.* **66** (Special Ed.), 177–188.

Hertz, H. G., Holz, M., Keller, G., Versmold, H., and Yoon, C. (1974). *Ber. Bunsenges. Phys. Chem.* **78**, 493–509.

Heubel, P.-H., and Popov, A. I. (1979). *J. Solution Chem.* **8**, 283–291.

Hill, A. V. (1910). *J. Physiol.* **40** (Proceedings, iv–vii).

Holz, M. (1978). *J. Chem. Soc. Faraday Trans. 1* **74**, 644–656.

Jackson, J. A., Lemons, J. F., and Taube, H. (1960). *J. Chem. Phys.* **32**, 553–555.

Kauffmann, E., Dye, J. L., Lehn, J.-M., and Popov, A. I. (1980). *J. Am. Chem. Soc.* **102**, 2274–2278.

Kessler, Y. M., Mishustin, A. I., and Podkovyrin, A. I. (1977). *J. Solution Chem.* **6**, 111–115.

Kintzinger, J. P., and Lehn, J. M. (1974). *J. Am. Chem. Soc.* **96**, 3313–3314.

Langer, H., and Hertz, H. G. (1977). *Ber. Bunsenges. Phys. Chem.* **81**, 478–490.

Langford, C. H., and Tong, J. P. K. (1977). *Acc. Chem. Res.* **10**, 258–264.

Laszlo, P. (1978). *Angew. Chim. Int. Engl. Ed.* **17**, 254–266.

Li, Z., and Popov, A. I. (1982). *J. Solution Chem.* **11**, 17–26.

Lindman, B., and Forsén, S. (1976). *In* "NMR Basic Principles and Progress" (P. Diehl, E. Fluck, and R. Kosfeld, eds.). Springer-Verlag, Berlin.

Meyer, F. K., Monnerat, A. R., Newman, K. E., and Merback, A. E. (1982). *Inorg. Chem.* **21**, 774–778.

Mishustin, A. I., and Kessler, Y. M. (1975). *J. Solution Chem.* **4**, 779–792.

Miura, K., and Fukui, H. (1980). *Inorg. Chem.* **19**, 995–997.

Popov, A. I. (1978). *In* "Proc. Symp. Spectrosc. Electrochem. Charact. Solute Species in Nonaqueous Solvents" (G. Mamantov, ed.) pp. 47–63. Plenum, New York.

Popov, A. I. (1979). *Pure Appl. Chem.* **51**, 101–110.

Popov, A. I., and Lehn, J.-M. (1979). *In* "Coordination Chemistry of Macrocyclic Compounds" (G. A. Melson, ed.), pp. 537–602. Plenum, New York.

Rose, K. D., and Bryant, R. G. (1979). *J. Magn. Reson.* **35**, 223–226.

Ruben, Y., and Rueben, J. (1976). *J. Phys. Chem.* **80**, 2394–2400.

Shih, J. S., and Popov, A. I. (1977). *Inorg. Nucl. Chem. Lett.* **13**, 105–110.

Silen, L. G., and Martell, A. E. (1964). "Stability Constants of Metal Ion Complexes," Special Publication No. 17. London Chemical Society, Burlington House, London.

Sisley, M. J., Yano, Y., and Swaddle, T. W. (1982). *Inorg. Chem.* **21**, 1141–1145.

Solomon, I. (1955). *Phys. Rev.* **99**, 559–565.

Swift, T. J., and Connick, R. E. (1962). *J. Chem. Phys.* **37**, 307–320.

Takahashi, A. (1977). *J. Phys. Soc. Jpn.* **43**, 968–975, 976–984.

Valiev, K. A. (1960). *Sov. Phys. JETP* **10**, 77–82; **11**, 883–890.

Valiev, K. A. (1962). *J. Struct. Chem.* **3**, 630–638.

Valiev, K. A. (1964). *J. Struct. Chem.* **5**, 477–488.

Valiev, K. A., and Khabibullin, B. M. (1961). *Russ. J. Phys. Chem.* **35**, 1118–1123.

Valiev, K. A., and Zapipov, M. M. (1966). *J. Struct. Chem.* **7**, 470–478.

Vogrin, F. J., and Malinowski, E. R. (1975). *J. Am. Chem. Soc.* **97**, 4876–4879.

Wehrli, F. W., and Wehrli, S. (1981). *J. Magn. Reson.* **44**, 197–207.

Weingärtner, H. (1980). *J. Magn. Reson.* **41**, 74–87.

7 Calcium-Binding Proteins

Hans J. Vogel
Torbjörn Drakenberg
Sture Forsén

Department of Physical Chemistry 2
University of Lund
Lund, Sweden

I. Introduction

The bivalent cations Ca^{2+} and Mg^{2+} differ markedly in their distribution in biological specimens. Most cells contain specialized pumps that utilize adenosine triphosphate (ATP) to excrete Ca^{2+} actively from the inside of the cell to the outside milieu. As a result, the intracellular calcium levels are kept as low as 10^{-7}–10^{-8} M, whereas the extracellular calcium levels

157

are usually on the order of 10^{-3} M (Kretsinger 1976, 1980). On the other hand, the influx of calcium into a cell (as may occur after stimulation by hormones or nerve impulses) provides a trigger par excellence for metabolic and mechanistic processes. Immediately after the stimulation is turned off, the calcium influx stops and calcium ATPases again reduce the intracellular levels of the cation, thus preparing the cell for another response (McLennan and Holland, 1975).

In contrast to calcium, Mg^{2+} ions seem to be much more evenly distributed throughout biological tissues. Estimates of this ion's total intracellular concentration are all in the range 1–5 mM, making it highly unlikely that significant short-term fluctuations can occur and have major effects on the activity of enzymes. This does not mean that Mg^{2+} ions have no role in the intracellular environment, but unlike calcium they do not act as intracellular second messengers. A large portion of intracellular Mg^{2+} is tightly bound in a 1 : 1 complex with ATP. This complex, rather than ATP itself, is the substrate in all enzymatic reactions using or producing this nucleotide (Mildvan, 1979).

Estimates of the intracellular metal ion concentration should ideally refer to the concentration of free ions rather than to the total metal ion concentration. Measurements of intracellular ion content are not easily performed. One of the best methods for Ca^{2+} determination relies on the calcium-dependent production of light by the protein aequorin when injected into a cell (Kretsinger, 1976). Estimates of internal Mg^{2+} concentrations can be based on the chemical shift differences measured in the three-line ^{31}PNMR spectra obtained from living tissue (Gupta et al., 1978). However, the interpretation of such shifts remains a debated subject (Gadian and Radda, 1981; Vogel and Bridger, 1982a; Wu et al., 1981). Although it should be theoretically feasible to estimate free metal ion concentrations directly by ^{25}Mg and ^{43}Ca NMR, sensitivity, as well as interpretation problems (see discussion on ^{23}Na NMR in living tissue, Civan and Shporer, 1978) renders this possibility quite unlikely.

Because nature has created two functionally different levels of calcium ions, it is not surprising that calcium-binding proteins can be divided into two groups, depending on the concentration of calcium ions required to saturate the metal-binding sites on the protein. Most extracellular proteins have dissociation constants on the order of 1 mM, whereas intracellular enzymes are generally sensitive to concentrations on the order of 10^{-5} to 10^{-8} M. Certain cytoplasmic enzymes have been found that appear to be sensitive to higher concentrations of calcium ions; however, the physiological significance of such observations seems rather obscure. For example, Ca^{2+}-dependent proteases have recently received a lot of attention, but millimolar concentrations of calcium were required for the activation of these intracellular enzymes. Only recently have proteases been discovered that respond to micromolar concentrations of calcium (Waxman, 1981). Another factor should be mentioned here: ATP can complex Ca^{2+} and Mg^{2+} with comparable affinity (Cohn and Hughes, 1962). However, the Ca^{2+}–ATP complex is usually less active as a substrate for phosphoryl-transferring enzymes (Mildvan, 1979). The ATP concentrations found in living tissue (Gadian and Radda, 1981) would be sufficient to sequester all intracellular calcium. This sort of restriction puts an upper limit on the amount of Ca^{2+} allowed to enter the cell and dictates the capacity of a possible buffer system and the calcium ATPases required for removal of the Ca^{2+} ion.

Intracellular Ca levels act as a cytoplasmic second messenger for hormonal signals or nerve impulses to a cell. Recent research on eukaryotic organisms has revealed that it is usually not the Ca^{2+} ion itself that redirects mechanistic and metabolic pathways but rather the calcium–calmodulin complex. This ubiquitous, highly conserved protein can act as an activator of many enzymes after it has bound calcium ions (Cheung, 1980, 1982; Klee *et al.*, 1980). Its properties will be discussed later. Nature has devised a second way of transmitting hormonal messages intracellularly via the second messenger, 3'5'-cyclic adenosine monophosphate (AMP). This compound can turn on protein kinases and slow down protein phosphatases in the cell, starting a cascade that results in the phosphorylation of several strategically chosen enzymes of intermediary metabolism and thus in modulation of the metabolic and sometimes mechanistic activity of cells (Krebs and Beavo, 1979; Perry, 1979). Both systems of hormonal control (namely, Ca^{2+}- and 3'5'-cyclic AMP-mediated) do not act independently. For example, Ca^{2+} (via calmodulin) affects the activity of the phosphodiesterase- as well as the hormone-stimulated membrane-bound enzyme adenylate cyclase which is responsible for the degradation and synthesis of the cyclic nucleotide. In addition, the calcium–calmodulin complex also interferes with the activities of phosphorylase kinase (Cohen, 1980) and of a protein phosphatase (Stewart *et al.*, 1982), thus providing direct calcium control over glycogen synthesis and breakdown, which is important for energy supply in muscle cells.

The role of calcium in muscle contraction is the best documented example of calcium control in a biological system (for reviews see Adelstein and Eisenberg, 1980; McCubbin and Kay, 1980; Perry, 1979). In striated muscle, Ca^{2+} ions bind to the troponin complex, inducing conformational changes that ultimately cause an interaction between the proteins actin and myosin, resulting in muscle contraction. In the less abundant smooth muscle cells contraction is stimulated by phosphorylation of one of the subunits of the contractile protein myosin called the myosin light chain.

The enzyme myosin light-chain kinase is fully dependent on Ca^{2+} and calmodulin for activity. This again illustrates the interplay between calcium–calmodulin and phosphorylation as a means of regulating biological activity.

II. An Overview of Calcium-Binding Proteins

Close to 100 different proteins have been reported to bind calcium ions. The binding of calcium can exert different effects on the protein. On the basis of these effects, calcium-binding proteins can be subdivided into five different classes.

The first group consists of a very few enzymes known to bind Ca^{2+} at the active site and to involve it in active catalysis. X-ray analysis of the crystal structures of phospholipase A_2 (Dijkstra et al., 1981) and *Staphylococcus* nuclease revealed a binding site for the metal ion at the active site; for the latter enzyme it is possibly involved in stabilizing binding of the substrate (Cotton and Hasen, 1971). For the nuclease, it has been reported that Mg^{2+} can bind to the calcium-binding site as well, but under these conditions the enzyme appears to be inactive. The nuclease is an exception in this respect because, as indicated earlier, most other phosphoryl-transferring enzymes have an almost absolute requirement for Mg^{2+} ions at their active sites (Mildvan, 1979). Mn^{2+} can usually substitute for Mg^{2+}, but replacement with Ca^{2+} may lead to inactivation. Although strictly speaking these enzymes could be classified as calcium-binding proteins, we shall not consider them as such, since the calcium binding has no apparent physiological significance.

A second class of calcium-binding proteins are the so-called modulator proteins. They modulate the activity of other proteins and enzymes. The best known examples of such modulators are troponin C (TnC) and calmodulin, but it is very likely that the proteins parvalbumin and S100 are also members of this group. Troponin C is part of the troponin complex, a 1 : 1 : 1 composite of the three proteins troponin T (TnT), troponin I (TnI), and TnC. The binding of calcium to TnC causes a large conformational change in this protein, which alters its interaction with the other two components, resulting in a release of the inhibition of the actin and myosin interaction responsible for contraction (Perry, 1979; McCubbin and Kay, 1980). Calmodulin also undergoes a large conformational change upon calcium binding. This exposes a hydrophobic surface that appears to be involved in the interaction of calmodulin with its target enzymes and proteins (LaPorte *et al.*, 1980; Marshak *et al.*, 1981; Klee *et al.*, 1980).

The third group consists of various enzymes markedly activated by the addition of calcium ions. A good example of such an enzyme is phosphoenolpyruvate carboxykinase from *Escherichia coli*. This enzyme requires Mg^{2+} at its active site for activity. It has, in addition, a site remote from the active site that is specific for calcium. Occupation of this site by calcium ions leads to marked activation of the enzyme (Goldie and Sanwal, 1980). The enzyme DNase I, purified from bovine pancreas (functioning in the intestines), may be activated by Ca^{2+} ions (Liao *et al.*, 1982). Similarly the calcium-stimulated proteases mentioned previously are members of this class (Waxman, 1981). Various other proteins activated by the addition of calcium act via calmodulin; hence the calcium-binding properties of such systems reside in the calmodulin molecule or sometimes even in the TnC molecule (Cohen, 1980).

The fourth group comprises various proteins involved in blood coagulation. Many of these proteins seem to bind calcium ions in a complex interplay with phospholipids (Stenflo and Suttie, 1977). Their localization and required calcium ion concentration for activation indicate that they are extracellular enzymes. A salient feature of these proteins is that all of them utilize the unusual γ-carboxyglutamic acid (Gla) residue in binding calcium. These Gla residues are posttranslationally carboxylated by a vitamin K-stimulated process (Stenflo and Suttie, 1977). Only one other protein not involved in blood coagulation has been described that also contains Gla residues. This protein is called osteocalcin and is localized in bone tissue (Hauschka and Carr, 1982).

The last group of enzymes includes those in which the binding of calcium is rather specific but where a biological function is not really known. Serine proteases such as trypsin and thermolysin bind Ca^{2+} without noticeable changes in the activity but with some changes in the stability of the enzyme (Epstein *et al.*, 1977; Matthews and Weaver, 1974). Another protein with a high affinity for calcium ions is the vitamin D-dependent intestinal calcium-binding protein, possibly involved in calcium uptake from the intestine (Wassermann, 1980). Osteocalcin, discussed earlier, is probably involved in the deposition of calcium in the bone. Calcium deposition could also be the function of the phosphodentine proteins purified from rat incisor dentine (Linde *et al.*, 1980; Zanetti *et al.*, 1981) or of the metal-binding proline-rich proteins purified from human saliva (Bennick *et al.*, 1981). Calcium binding to the hen egg yolk phosphoprotein has been reported, but its function remains unclear (Grizutti and Perlman, 1975). Actin, best known as a major contractile protein in muscle cells, but highly abundant in all eukaryotic cells, contains one tightly bound ATP nucleotide as well as one Ca^{2+} ion (Clarke and Spudich, 1977). Concanavalin A purified from jack bean contains one Mn^{2+}- and one Ca^{2+}-binding site (Becker *et al.*, 1975).

III. Structure of the Calcium-Binding Site

X-ray crystallographic studies on six different calcium-binding proteins have indicated that only oxygen ligands and water molecules are involved in the complexation of calcium ions. This observation has been largely supported by ^{113}Cd-NMR studies with chemical shifts compatible with this notion, as discussed later. The structure of the binding site as generally observed suggests octahedral coordination of the metal ion (Kretsinger, 1976). No radically different forms of ligation have been observed as yet in biochemical systems, although other forms have been reported for small model compounds. Since negatively charged or uncharged oxygen atoms are responsible for metal ion binding, only certain amino acids participate in the complexation process. The side chains of the amino acids aspartic acid and glutamic acid are often involved in ligand formation, providing an explanation for the usually very acidic nature of these proteins. Sometimes both oxygen atoms function in concert as one ligand [as in phospholipase A_2 (Dijkstra *et al.*, 1981 (Fig. 1)]. A second major contributor to the ligands comprising a calcium-binding site are carbonyl functions from the protein's backbone. Also, hydroxyl groups of Ser, Thr, and Tyr residues may serve as ligands.

In addition to these types of ligands, a few less common amino acids have been proposed to be involved. Dianionic phosphoryl moieties such as those found in the amino acid phosphoserine (which is part of the proteins purified from dentine), in egg yolk phosvitin, in egg white ovalbumin, in human saliva proteins, and in the milk protein casein are often

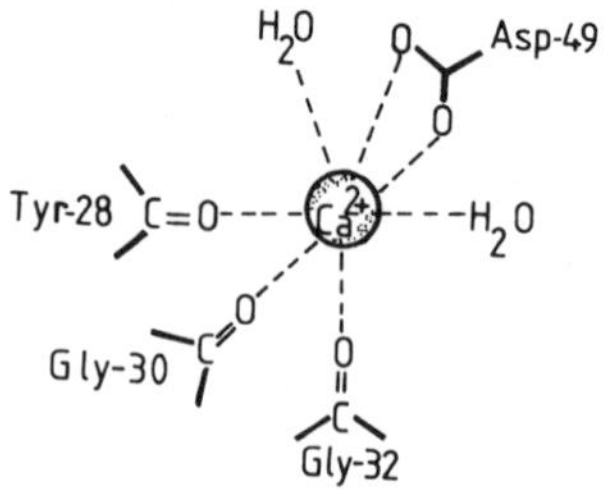

Fig. 1 A schematic representation of the calcium-binding site in bovine pancreas phospholipase A_2 (from Dijkstra *et al.*, 1981).

thought to be important for calcium ion complexation. Phosphatase treatment of these proteins can remove the phosphorus moieties, thus leading to a lower metal ion binding affinity (Bennick *et al.*, 1981). Also of importance is the recently discovered amino acid γ-carboxyglutamic acid previously mentioned. It is noteworthy that the Gla residues are very unevenly distributed throughout the amino acid sequence. For example, in prothrombin all 10 Gla residues and most calcium-binding activity are contained in the first 40 amino acids. This peptide is highly homologous with Gla-containing peptides purified from the other proteins involved in blood clotting (Stenflo and Suttie, 1977).

Knowing now which amino acids are involved, we can take a closer look at the structure of the actual binding sites. Inspection of the crystal structure of various proteins reveals conformations in which oxygen atoms from various amino acid residues may form a binding pocket for a Ca^{2+} ion. For example, in *Staphylococcus* nuclease residues 19 and 21, as well as residues 40, 41, and 43, contribute to the binding site (Cotton and Hazen, 1971). In phospholipase, backbone carbonyl functions from Tyr-28, Gly-30, and Gly-32, the carboxylate oxygens of Asp-49, and two water molecules serve as ligands for the calcium ion (Dijkstra *et al.*, 1981) (Fig. 1). Similar sites exist in trypsin and thermolysin. They differ from the EF-hand structure where all the amino acids involved at a single calcium-binding site are contained in one small stretch of amino acids.

Calcium-binding sites containing Gla residues may derive their affinity from the fact that the side chain of one Gla residue can accommodate half of the ligand requirements for a calcium ion. Two Gla residues opposite each other are sufficient to comprise a high-affinity site (Furie *et al.*, 1979). A completely different mode of metal binding has been proposed for two calcium-binding proteins with a high charge density—egg yolk phosvitin and the phosphodentine protein. It has been suggested that for either protein the metal ions can actually diffuse over the negatively charged protein surface (Cookson *et al.*, 1980, Vogel, 1983).

One type of binding that has attracted a large amount of attention involves the calcium-binding site originally observed in the X-ray structure of parvalbumin. Elucidation of the crystal structure of this protein revealed the presence of six α-helices labeled A through F. The loops between the CD and EF helices form two tight calcium-binding sites with an octahedral arrangement (Kretsinger and Nockolds, 1973). As can be seen in Fig. 2, the loop between the two helices is wrapped around the calcium ion in such a way that the ligand sites are filled in the order XYZ−Y−X−Z. The −Y ligand at such sites is always a carbonyl oxygen from the protein's backbone, whereas all the other ligands are amino acid side chains (mainly Asp and Glu residues) and occasionally a water mole-

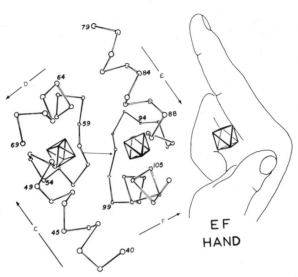

Fig. 2 CD and EF regions of parvalbumin and the EF or Kretsinger hand showing the orientation of the helices (from Kretsinger, 1976, reproduced with permission).

cule (Kretsinger, 1976, 1980). It is obvious that such an arrangement provides for a regular and easily recognizable amino acid sequence. Alignment of the amino acid sequences of a series of calcium-binding proteins showed that similar structures were present in skeletal and cardiac TnC (four and three sites, respectively), calmodulin (four sites), the S100 subunits (one site each), the light chains from skeletal muscle myosin (four sites), the vitamin D-dependent intestinal calcium-binding protein (two sites), and possibly T_4 lysozyme (Isobe and Okuyawa, 1978, 1981; Kretsinger, 1976, 1980; Tufty and Kretsinger, 1975). On the basis of the sequence homology with parvalbumin, a model for the TnC structure has been computed (Kretsinger and Barry, 1975). This prediction awaits experimental confirmation by X-ray diffraction studies. Such a comparison between predicted and determined structure has been possible in the case of the vitamin D-dependent calcium-binding protein (Szebenyi et al., 1981). Fortunately, the crystal structure showed indeed the existence of the two helix–loop–helix binding domains predicted on the basis of the amino acid sequences. The structure of one site is very similar to the EF hand of parvalbumin, the other differs in some details (Szebenyi et al., 1981).

Because of the importance of this EF-hand arrangement for the ubiquitous modulator proteins TnC and calmodulin, many studies have tried to clarify the high calcium affinity of this structural element. Three contributing factors have been identified: (1) numerous studies with proteolytic, chemically cleaved, or synthetic peptides representing parts of parvalbumin, skeletal TnC, or calmodulin have indicated that placement of amino acids in the right sequences is not a sufficient condition for calcium binding. However, such a sequence is required, because parvalbumin and cardiac TnC each contain one site with amino acid substitutions that is thought to have the right structure but does not bind calcium. Moreover, myosin light chains have four homologous EF-hand-like loops, but do not bind calcium after purification. Hence other features must play a role in determining the calcium affinity of such sites.

(2) When peptide fragments with helix–loop–helix arrangements were studied, moderate calcium affinity was commonly found (for references, see the following discussion). One possible explanation may be offered here to explain why a helix–loop–helix arrangement provides higher calcium-binding affinities than the loop alone. Because of the accumulation of individual amino acid dipoles, helices give rise to quite considerable electrical dipoles. The N-terminal end of such a helix dipole can bind negatively charged phosphate ions (Hol *et al.*, 1978). Likewise, positively charged ions, such as Ca^{2+}, can be stabilized at the C-terminal end of an α-helix. Such forces could explain why no direct relation exists between the binding strength and the number of ligands (Kretsinger, 1976). We feel that this possibility deserves further investigation. Other Chou–Fasman-type calculations have suggested that β turns in the calcium loop contribute to the affinity (Vogt *et al.*, 1979).

(3) Only proteolytic fragments containing two calcium-binding sites (e.g., helix–loop–helix–helix–loop–helix) closely resemble native proteins with respect to calcium-binding properties (parvalbumin: Coffee and Solano, 1976; Derancourt *et al.*, 1978; Maximov and Mitin, 1979; TnC: Gariépy *et al.*, 1982; Grabarek *et al.*, 1981; Leavis *et al.*, 1978; Reid *et al.*, 1981; calmodulin: Andersson *et al.*, 1983b; Drabikowski *et al.*, 1977; Vogel *et al.*, 1983; Walsh *et al.*, 1977). Thus the structural relation between the two sites is important in addition to the specific amino acid sequence and the helices on either side of the calcium-binding loop. The crystal structure of the intestinal calcium-binding protein revealed a structural relationship between the two sites similar to that in parvalbumin (Szebenyi *et al.*, 1981). Moreover, chemical modification of the single Arg-75 residue in parvalbumin results in a large decrease in the calcium-binding constants (Coffee and Solano, 1976). This residue is far (>20 Å) from the calcium-binding sites. It forms a salt linkage with the Glu-81 residue and as such is responsible for preserving the relationship between the two calcium-binding loops.

The concept that the structural relationship between the two sites is preserved leads to the prediction that calcium-binding proteins must be structurally rigid to attain high calcium affinity. Other evidence, however, does not support such models. Studies with spectroscopic techniques have indicated that large conformational changes occur upon binding of calcium. Particularly, circular dichroism studies have been illustrative in this respect, since they have revealed a large increase in ellipticity upon calcium binding, suggesting increases in the α-helix content of these proteins (McCubbin and Kay, 1980). Studies with synthetic peptides have further demonstrated that calcium-induced folding of α-helices around the calcium-binding loop may occur (Reid et al., 1981). These results indicate a flexibility of the calcium-binding sites and other areas of the calcium-binding proteins. To shed further light on these questions, NMR studies will be of help.

To distinguish between rigid and flexible calcium-binding sites one can determine the dynamics of the bound metal ion. In addition, determination of the exchange rates of the metal ion is important, because it can be intuitively expected that on-rates comparable to the diffusion-limited rates will occur for very flexible sites, whereas rigid sites are expected to give rise to hindered access resulting in lower on-rates. Both types of information have been obtained from direct metal NMR studies and will be considered in the following discussion.

Further experiments should involve studies on the dynamics of the individual amino acid residues in the protein. Numerous ^{1}H-NMR studies have been reported for these proteins, but all suffer from the fact that only a few resonances are assigned and that relaxation measurements (because of effective cross relaxation) are not a direct measure of the dynamics of the protein (Kalk and Berendsen, 1976; Sykes et al., 1978). High-resolution ^{13}C NMR holds promise in this respect, but the low natural abundance of this nucleus combined with its poor sensitivity makes the amounts of protein required for such studies quite prohibitive. However, early studies concerning the dynamics of parvalbumin as well as studies on the lanthanide-induced broadening in a prothrombin fragment indicate the potential of the technique (Nelson et al., 1976, Furie et al., 1979).

IV. Nuclear Magnetic Resonance Studies on Calcium-Binding Proteins

In this section we shall discuss recent NMR studies reported for calcium-binding proteins. Several high-resolution ^{1}H-NMR studies have been reported, but they fall outside the scope of this chapter. We shall focus on [23]Na-, [25]Mg-, [43]Ca-, and [113]Cd-NMR studies on calcium-binding proteins.

Where necessary, comparisons with results obtained by other spectroscopic techniques will be discussed. Studies on quadrupolar nuclei binding to other proteins have been reviewed elsewhere (Forsén and Lindman, 1981).

A. Small Peptides

Ion NMR studies on metal binding to small peptides hold great promise,
since the low molecular weights ensure that measurements are done in the
extreme narrowing limit ($\omega\tau_c \ll 1$) and the relaxation data are easily interpreted. Yet only a few studies have been reported. A dipeptide consisting
of two Gla residues has been studied to evaluate the usefulness of ^{43}Ca
and ^{25}Mg NMR in the study of Gla-containing calcium-binding proteins.
Addition of the peptide to solutions of enriched ^{25}Mg^{2+} resulted in drastic
increases in the measured line width (Fig. 3a). No changes in the chemical
shift position were detected, consistent with our data on a variety of small
model compounds (unpublished observations). Based on a fast exchange
between free and bound Mg^{2+}, the dissociation constant was estimated to
be 0.6 mM by Robertson et $al.$, (1978). These authors comment that,

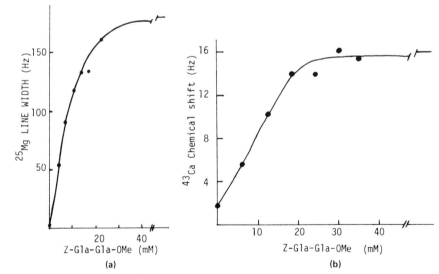

Fig. 3 (a) The observed ^{25}Mg line width as a function of an added peptide, Z-D-Gla-D-
Gla-OMe. The magnesium concentration varied from 21.3 to 18.7 mM at pH 6.5–6.6 and $t =$
27°C. (b) The ^{43}Ca-NMR chemical shift as a function of an added peptide, Z-D-Gla-D-Gla-
OMe. The calcium concentration varied from 19.5 to 17.4 mM at pH 6.6 and $t =$ 27°C (from
Robertson et $al.$, 1978, reproduced with permission).

because of the large quadrupole moment for this nucleus (0.22×10^{-28} m^2), binding to macromolecules would broaden out the lines beyond detectability. ^{43}Ca-NMR studies showed only small changes in line width upon addition of the dipeptide, probably because of the relatively small quadrupole moment for this nucleus (0.06). However, changes in chemical shift were observed (Fig. 3b), leading to an estimate of the dissociation constant for calcium of 0.6 mM. This is consistent with reported shifts for other small molecules (Drakenberg, 1982, unpublished observations). Competition studies reported by Robertson et al. (1978) showed that Mg^{2+} and Ca^{2+} competed for the same site. ^{25}Mg-NMR studies reported for similar peptides revealed identical binding constants for the Mg^{2+} ion, whereas a pK_a of 4.8 was determined for the disappearance of the excess magnesium line width at low pH values. These authors suggest that this reflects the pK_a of the γ-carboxylglutamic acid groups (Robertson et al., 1979). However, this behavior seems somewhat unusual considering the proximity of the two carboxylate groups. For example, in malonate [$CH_2(COOH)_2$], the titration of the second carboxylate group is at a higher pH, because of the deprotonation of the first. These studies illustrate the applicability of the method based on an excess of metal ion in the study of weak binding sites ($K_d > 0.1$ mM). Similar results can be anticipated from studies on proteins with low-affinity binding sites.

B. Prothrombin

Prothrombin is one of the vitamin K-dependent zymogens involved in the blood coagulation process. (Stenflo and Suttie, 1977). The N-terminal peptide called fragment 1 (MW 23,500; residues 1–156) can be proteolytically removed and contains all 10 Gla residues in its first 45 amino acids as well as the calcium-binding capacity and the Ca ion-dependent phospholipid binding capacity of the zymogen. Both calcium and phospholipid binding are important in the activation of prothrombin, the ultimate conversion in the proteolytic cascade leading to the formation of fibrin clots from fibrinogen (Stenflo and Suttie, 1977). When carboxylation of the Glu residues to Gla is inhibited, the resulting prothrombin has a very low affinity for calcium ions. ^{13}C-NMR studies on a subfragment (residues 12–44) of prothrombin fragment 1, containing eight Gla residues, showed one high-affinity metal-binding site (0.5 μM) for the binding of paramagnetic gadolinium ions as calcium substitutes. Reduction and alkylation of the internal disulfide bond (residues 18–23) abolished the high-affinity site. A model was proposed in which two Gla residues are placed opposite each other, thus constituting one high-affinity calcium-binding site (Furie *et al.*, 1979).

The first ^{25}Mg-NMR studies on bovine prothrombin fragment 1 were performed with a 100-fold excess of Mg^{2+} over fragment 1. Two groups, one with $pK_a = 4.5$ and one with $pK_a > 7.0$, seem to determine Mg binding. Hence side chains other than those contributed by the Gla unit may be involved in metal ion binding by prothrombin fragment 1 (Robertson *et al.*, 1979). In subsequent studies reported by Marsh *et al.* (1979), ^{43}Ca as well as ^{25}Mg NMR was used. All experiments were performed in the presence of 100 m*M* NaCl to avoid nonspecific weak binding of divalent metal ions to the protein. Studies with pH titration and metal competition revealed the presence of several sites with differing behavior. Ca^{2+} binds in general more strongly than Mg^{2+}. Here one class of sites is accessible only to Mg^{2+} and not to Ca^{2+}. When sufficient Ca^{2+} is present, Mg^{2+} binding is dependent only on a pK_a of 4.5, suggesting that the sites characterized by a $pK_a > 7.0$ are most important for calcium binding.

The dependence on $pK_a > 7$ seems unusual for metal ion binding, since normally only carboxylate groups ($pK_a < 6$) are important. It may reflect titration of a histidine residue, as recently determined by ^{1}H NMR (Pletcher *et al.*, 1981). The latter study also drew attention to possible protein aggregation, especially below pH 5, which has not been considered as an additional parameter causing line broadening in metal NMR studies. It should be emphasized again that ^{43}Ca- and ^{25}Mg-NMR studies are performed with a large excess of metal ions over binding sites. The NMR line widths are not easy to interpret because contributions from intermediate-exchanging sites, fast-exchanging sites, and free metal ion are lumped together into a single spectral line and only average parameters for these sites are obtained. Under the experimental conditions used, resonances for slow-exchanging protein-bound calcium ions would be overlooked. Experience with other proteins in our laboratory indicates that for a protein of this size such a line would be ~1000 Hz wide. This is not easy to detect reliably on most low-field spectrometers because of the low resonance frequencies of the Ca and Mg nuclei, causing the "ring-down" from the probe to show up in the spectra.

To circumvent this problem, two other methods are open to the experimentalist who wishes to explore the properties of strong calcium-binding sites. One possibility is the study of ^{23}Na NMR. The Na^+ ion (0.098 nm) has a radius very similar to that of a calcium ion (0.099 nm), a high resonance frequency, and favorable receptivity (Detellier, Chapter 5, second volume). The lower charge of the sodium ion results in weaker binding of this nucleus. Strong calcium-binding sites can be easily studied by competition experiments between Ca^{2+} and Na^+. The second alternative is the use of a nucleus with spin $\frac{1}{2}$. $^{113}Ca^{2+}$ has the same charge as Ca^{2+} and a virtually identical ionic radius. Studies in our laboratory have shown that cadmium ions readily substitute for calcium ions in a large number of calcium-binding proteins. Although its affinity is not always identical to that of calcium, extensive characterizations by ^{1}H-NMR and two-dimensional coupling correlation ^{1}H-NMR spectroscopy indicate that calcium-binding proteins with EF-hand calcium-binding sites adopt similar conformations on substitution with Ca^{2+} or Cd^{2+} (Ragg et al., 1984).

C. Osteocalcin

Osteocalcin is an abundant, small calcium-binding protein of 49 amino acids (MW 5500) localized in the bone matrix. This protein adheres strongly to hydroxylapatite. It contains three Gla residues and binds calcium ions with an affinity of about 0.8 mM (Hauschka and Gallop, 1977). This dissociation constant compares very favorably to those measured earlier for the Gla-containing small peptides (Robertson et al., 1978, 1979). Recent ^{43}Ca-NMR experiments in our laboratory have revealed a similar binding constant for the calcium ion (Andersson, Swärd and Drakenberg, unpublished observations): On binding of calcium, the protein undergoes a rather large increase in α-helical content (Hauschka and Carr, 1982).

D. Hemocyanin

Hemocyanins are large, copper-containing proteins of molluscs and arthropods involved in oxygen binding and transport. The protein from the spiny lobster *Panuliris interruptus* is a hexamer of molecular weight 4.5×10^5. Ca^{2+}, Mg^{2+}, and Na^+ ions have been known for some time to exert effects on the structure and biological function of these proteins (Antonini and Chiancone, 1977). Initial ^{23}Na-NMR studies on the protein revealed the existence of two classes of binding sites with differing affinities for the metal ion. Calcium displaced only the more tightly bound Na^+ ions, suggesting competition for the same sites (Norne et al., 1979). Oxyhemocyanin had a higher affinity for the Na^+ ions than deoxyhemocyanin. More recent studies applied ^{43}Ca as well as ^{23}Na NMR. ^{43}Ca NMR turned out to be the most sensitive technique for probing the properties of the sites with low calcium affinity. However, ^{23}Na-NMR studies revealed the presence of one strong calcium-binding site which was undetectable by

^{43}Ca NMR because of the high molecular weight of the protein (Andersson *et al., 1982b*). From the ^{43}Ca-NMR and ^{23}Na-NMR data it was concluded that the calcium ion had considerable mobility when bound to the protein.

E. Trypsin

Trypsin, a well-known serine protease (MW 24,000), can bind one calcium ion with a dissociation constant of about 0.03 mM (Cliffe and Grant, 1981). Thus this site is likely to be filled in the intestine where the protein resides. No changes in activity have been observed on binding of calcium, but the protein is markedly stabilized against self-degradation in the presence of calcium ions (Epstein *et al.*, 1977). The bound calcium ion has six ligands: two carboxyl groups, two carbonyl groups, and two water molecules (Kretsinger, 1976). Although the binding is not very strong, preliminary observations in our laboratory have shown that free and protein-bound calcium exchange slowly (Drakenberg and Vogel, 1983). The protein-bound ^{43}Ca resonance is shifted 10 ppm downfield from the resonance of the free hydrated ion. Also, one slow exchanging site was observed in ^{113}Cd-NMR spectra. Thus far no changes have been observed in this resonance on binding of dipeptide or proteinous trypsin inhibitors to the active site. The observation of resonances in the slow-exchange limit suggest an on-rate that is much less than the diffusion limit for the calcium ion. Hence Ca^{2+} access to its site may be sterically hindered.

F. Thermolysin

Thermolysin is an enzyme with proteolytic activity secreted by *Bacillus* strains and, like trypsin, is stabilized by Ca^{2+} ions. In this case four calcium ions are bound to the protein in addition to the catalytic zinc residue (Matthews and Weaver, 1974). All four calcium sites are six-coordinate (Kretsinger, 1976). Unfortunately just one rather unclear report of ^{43}Ca- and ^{67}Zn-NMR studies on this interesting enzyme has appeared (Shimizu *et al.*, 1982). Although the authors used a large excess of metal ion over protein, they referred to the only relatively narrow resonance present in the ^{43}Ca and ^{67}Zn spectra as that of the metal–protein complex. It seems more likely to us that this resonance represents the somewhat exchange-broadened resonance of the remaining free metal ion, whereas the protein-bound resonance is so broad that it contributes to the baseline distortions clearly visible in all the spectra.

G. Phosphorylated Proteins

For the human saliva protein ^{31}P-NMR pH titration studies in the absence and presence of metal ion provided direct evidence of the importance of the phosphoserine side chains in metal ion complexation (Bennick *et al.*, 1981). Chemical shift changes measured for the two phosphoserine residues of hen egg white ovalbumin on metal ion addition did not support the suggestion that the proteins phosphate groups are involved in metal ion complexation *in vivo* (Vogel and Bridger, 1982b), and the outcome of preliminary ^{25}Mg-NMR experiments seems to agree with these results. Other proteins with phosphoserine residues involved in Ca^{2+} and Mg^{2+} binding are phosvitin from hen egg yolk (Grizutti and Perlmann, 1975) and the phosphodentine proteins from teeth (Linde *et al.*, 1980). Various NMR studies on the latter have demonstrated its metal binding properties. It was suggested that the cations can diffuse along the protein surface and form a relatively long-lived metal–phosphoprotein complex (Cookson *et al.*, 1980). A similar model has been proposed for phosvitin (Vogel, 1983). Studies on ion binding to these proteins using quadrupolar nuclei are currently in progress in our laboratory, and measurements of the relaxation properties of the metal ions bound to the proteins should allow a test for the proposed model.

H. Phospholipase A$_2$

Phospholipase A$_2$ (MW 14,000) hydrolyzes phospholipids in micellar aggregates. For optimal activity, binding of Ca^{2+} ions to the active site of the enzyme is necessary (Verheij *et al.*, 1980). The binding site for the calcium appears octahedral, but both ligands of the carboxylate group of Asp-49 residue contribute, so a total of seven oxygen atoms may interact with the metal ion (Fig. 1). The enzyme also exists in a zymogen form which is converted to the active enzyme by the tryptic release of an N-terminal heptapeptide. Both the zymogen and the active enzyme have been studied by our group with the ^{43}Ca-NMR technique (Andersson *et al.*, 1981a; Andersson, 1981). The zymogen binds only one Ca^{2+} ion with an affinity of 0.25 mM (Fig. 4a). An estimation of the off-rate from measurements of the temperature dependence of the line width, as detailed in Chapter 8, Vol. II on ^{43}Ca and ^{25}Mg NMR, resulted in an off-rate of 3.0 × 10^{-3} s^{-1}. Combined with the binding constant, this figure gives rise to an on-rate much less than the diffusion limit, indicating hindered access of calcium to its binding site (Forsén *et al.*, 1981). The pH dependence of the calcium line width indicates that a group with a pK_a of 5.2 is involved in

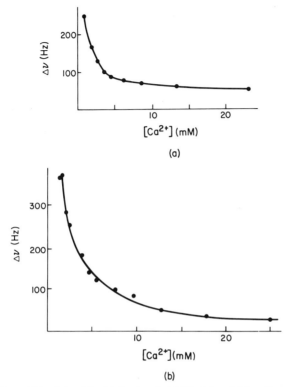

Fig. 4   The line width of the ⁴³Ca-NMR signal. (a) [PLA₂] = 0.74 mM at pH 7.1 and $t =$ 23°C. The solid line represents the theoretical curve calculated assuming two binding sites. $K_s = 10.10^3 \, M^{-1}$ and $K_w = 15 \, M^{-1}$. (b) [PPLA₂] = 2.0 mM at pH 7.5 and $t = 23$°C. The solid line represents the theoretical curve calculated assuming one binding site. $K = 2.5 \times 10^3 \, M^{-1}$ PLA₂, Phospholipase A₂ (from Andersson 1981).

the calcium binding. This is presumably the pK_a of the carboxylate group of Asp-49.

Subsequent studies on the active form of the enzyme revealed a slightly different picture. Two calcium-binding sites were discovered with affinities of 1 mM (active site) and a new site with 60 mM, respectively (deduced from the data in Fig. 4a). These values agree quite well with those reported by Slotboom *et al.* (1978). The off-rate for the strong site was estimated to be $1 \times 10^3 \, s^{-1}$, and the quadrupole coupling constant was estimated at 1.37 MHz, as mentioned later. By comparing pH titrations at two calcium/enzyme ratios it could be deduced that the strong site sensed a pK_a of 5.7, whereas the weak site was responsive to a pK_a of 4.6. This differing behavior can be clearly seen in the titration curves of Fig. 5.

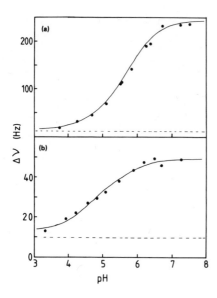

Fig. 5 The pH dependence of the ^{43}Ca-NMR line width at 23°C. (a) A 1.1 mM CaCl$_2$ solution containing 0.98 mM PLA$_2$, pK$_1$ = 5.7 (87%), pK$_2$ = 4.6 (13%); (b) a 11.5 mM CaCl$_2$ solution containing 1.1 mM PLA$_2$, pK$_1$ = 5.7 (30%), pK$_2$ = 4.6 (70%).

I. Insulin

Insulin can exist in solution as a torus-shaped hexamer with a total molecular weight of ~40,000 (Blundell *et al.*, 1972). The two Zn^{2+} sites are located 17 Å apart on the threefold symmetry axis. Each zinc ion is coordinated to three histidine nitrogens and three water molecules. X-ray studies on a crystal containing three Cd^{2+} ions per hexamer indicate that two of the Cd^{2+} ions bind at the Zn^{2+} sites, whereas the third is located in the middle of the hexamer close to the B-13 Glu residues. Sudmeier *et al.* (1981) have studied the insulin hexamer with ^{113}Cd NMR. The protein loaded with three ^{113}Cd^{2+} ions per hexamer gave rise to two ^{113}Cd-NMR signals with an intensity ratio of 2 : 1 and with chemical shifts of 165 and −36 ppm. The high-frequency signal disappeared after the addition of two equivalents of Zn^{2+} ions, whereas the −36-ppm signal decreased with increasing Ca^{2+} concentration. These experiments show that the insulin hexamer has both Zn^{2+}- and Ca^{2+}-binding sites and that Zn^{2+} ions do not bind strongly to Ca^{2+} sites, whereas Cd^{2+} binds to all three sites.

J. Concanavalin A

Concanavalin A (Con A) has a molecular weight of 25,500. Solutions of Con A consist of dimers and tetramers with a composition dependent on the medium. It is isolated from jack beans and has been extensively studied because it has unusual biological properties (Bittiger, 1977). Concanavalin A interacts with saccharides in a specific way only when each subunit contains two metal ions, at sites S1 and S2. In the native protein site S1 is occupied by Mn^{2+} and site S2 by Ca^{2+}. Both sites can also bind other metal ions, such as Cd^{2+}. Ellis and co-workers (Bailey *et al.*, 1978; Palmer *et al.*, 1980) have studied the metal binding to Con A with ^{113}Cd NMR. Concanavalin A can exist in two forms, locked or unlocked. Only the locked form gave rise to ^{113}Cd signals from protein-bound cadmium ions. Normally the ^{113}Cd-NMR spectrum of cadmium-loaded protein (two Cd^{2+} per monomer) showed three signals: 68, 46, and -125 ppm (Fig. 6). Various metal competition experiments were performed to assign the different signals to specific metal-binding sites. In this way it was possible to identify the 68-ppm signal as due to free cadmium, and the 46- and -125-ppm signals were assigned to the Mn^{2+} and Ca^{2+} sites, respectively (Fig. 6). The fact that a signal due to free $^{113}Cd^{2+}$ was always observed, even when only 1 equiv had been added, indicates that the binding of cadmium to the two sites is not very strong. Binding constants of 3.1×10^3 and 2.1×10^3 M^{-1} were derived for the S1 and S2 sites. The off-rates of cadmium ions from the protein are quite slow, as seen from the ^{113}Cd-NMR spectra, and this shows together with the low binding constants that the cadmium ion on-rates to the two sites are several orders of magnitude

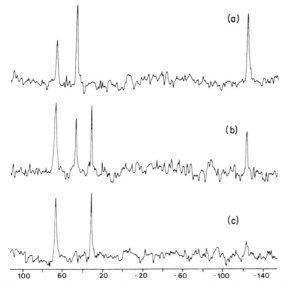

Fig. 6 ^{113}Cd-NMR spectra of Con A. (a) Con A containing 2.2 equiv of $^{113}CdCl_2$ per monomer; (b) sample (a) to which 1 equiv of Ca(11) has been added; (c) sample (a) to which 1 equiv of Mn(11) has been added. (Reproduced with permission from Palmer *et al.*, 1980. Copyright 1980 American Chemical Society.)

slower than the diffusion limit. This is similar to what has been found for trypsin and phospholipase A_2 but differs from the behavior of the EF-hand proteins parvalbumin, TnC, and calmodulin, now to be discussed.

K. Parvalbumin

Parvalbumins are small, highly soluble acidic proteins found in the cytoplasm of vertebrate skeletal muscles (MW 12,000). These proteins have a high affinity for calcium (10^{-8}–10^{-9} M), and they bind two calcium ions each in an EF-hand structure. Next to these two strong EF-hand-type sites, the protein can contain a third weaker metal-binding site (1 mM, as discussed later). Although the physiological role of parvalbumins is not explicitly known, it has been suggested that they act as a calcium buffer that relaxes the muscle by depleting TnC of calcium before it releases the cation to the sarcoplasmic reticulum (Gerday and Gillis, 1976; Pechère, 1977). This role seems compatible with their distribution in only fast muscle fibers (Celio and Heizmann, 1982) which are known to relax faster than slow muscle fibers which are devoid of parvalbumin.

Early ^{23}Na-NMR binding studies were reported by Laszlo's group (Grandjean et al., 1977). They compared the Na binding properties of fully calcium-loaded parvalbumin (PaCa$_2$) with that of parvalbumin depleted of one of its calcium ions (PaCa) by treatment with the complexing agent 1,2-di(aminoethoxy)ethane-N,N,N',N'-tetraacetate (EGTA). The addition of PaCa to a solution of NaCl results in a large increase in the observed line width, whereas the addition of PaCa$_2$ produces only minor effects. Because parvalbumins have isoelectric points at about pH 5.0, it is not possible to obtain accurate pH titration data. In competition experiments the following sequence of binding affinities was observed: Na$^+$ < K$^+$ ≪ Mg^{2+} < Ca^{2+} ≪ Mn^{2+}. In subsequent ^{23}Na-NMR studies it was alleged that EGTA would bind to parvalbumin and would thus interfere with the obtained results (Parello et al., 1979). It was shown that, for PaCa preparations prepared in the absence of a chelating agent, only weak Na$^+$ binding was observed. Hence it was suggested that the previously observed high-affinity site for Na$^+$ was in fact due to protein-bound EGTA. The binding of complexing agents to other calcium-binding proteins has been reported as well (Haiech et al., 1979). Be that as it may, we are inclined to the view that the reported experimental evidence for EGTA binding to parvalbumin is very poor and requires further investigation.

^{113}Cd-NMR experiments have been quite successful in the study of parvalbumin. Drakenberg et al. (1978) demonstrated that the binding of 2 equiv of Cd^{2+} resulted in two well-resolved resonances at -94 and -98 ppm, which were assigned to the Cd^{2+} bound to the CD and EF hands, respectively. These shifts indicate that only oxygen atoms are involved in the complexation. Nitrogen or sulfur ligands normally give rise to different shifts (see Armitage this series). We now believe that the original assignments were unfortunately wrong. Experiments with paramagnetic Gd^{3+} ions were used in the original assignment of the peaks, however, later experiments performed in our laboratory at higher frequencies showed that the earlier reported Gd^{3+}-induced disappearance of one of the signals (Drakenberg et al., 1978) was actually caused by broadening induced by Gd^{3+} bound to the weak third site, rather than by competition and displacement. On the basis of this and other arguments it appears that the original assignment should be reversed. The third site was originally discovered through observation of the Mg^{2+}-induced shifts for one of the two Cd^{2+} resonances, and its properties have been explored with ^{25}Mg NMR (Cavé et al., 1979). This site also binds univalent cations, is readily apparent only in the β series of parvalbumins (Cavé et al., 1982), and is located less than 7 Å away from one of the strong sites (Swärd et al., unpublished observations). Interesting results have been obtained with a variety of rather homologous but slightly different parvalbumins purified from different sources. Although two peaks are always observed, differences in chemical shifts are observed for the different species, and resonances for EF- and CD-hand-bound $^{113}Cd^{2+}$ may even change position (Cavé et al., 1982). These results illustrate the great sensitivity of cadmium shifts to minute changes in protein structure.

X-ray crystallographic studies on carp III parvalbumin revealed that the lanthanide ion terbium first replaced Ca^{2+} bound to the EF hand before replacing the calcium ion in the CD hand (Sowadski et al., 1978). We have studied replacement of the two Cd^{2+} ions by a variety of lanthanide ions using ^{113}Cd NMR. The results are extremely dependent on the nature of the different lanthanide ions. For example, addition of the diamagnetic lutecium ion (which has the smallest ionic radius) displaces selectively only the high-field resonance, whereas addition of the diamagnetic lanthanum ion (which has the largest ionic radius) replaces both cadmium ions simultaneously. Paramagnetic lanthanide ions of intermediate size affect the cadmium resonances in a systematic way related to their size (Drakenberg et al., unpublished results). These results are of paramount importance because many investigators have assumed that substitution of lanthanide ions for calcium ions bound to calcium-binding proteins is the same for the different lanthanides (For review see Martin and Richardson, 1979). For example, Horrocks and collaborators (1981) have used the luminescent properties of different lanthanides, whereas others have studied the paramagnetic shifts induced in the ^{1}H-NMR spectra of parvalbumin by replacing calcium with ytterbium (Lee and Sykes, 1981). Our results obviate the need for appropriate control experiments and suggest that greater care should be taken in the interpretation of studies using lanthanide ions as calcium ion replacements.

Recent ^{43}Ca-NMR studies in our laboratory have monitored directly the resonances of calcium ions tightly bound to parvalbumin (Fig. 7) (Andersson et al., 1982a). Measurements of the longitudinal relaxation rate indicated a R_1 of 1.6×10^3 s^{-1}. Combined with the R_2 estimated from the line width, a correlation time of 4 ns was derived for the calcium ion, assuming quadrupolar relaxation as the only relaxation mechanism. The quadrupolar coupling constant was estimated at 1.3 MHz. The overall correlation time for this protein as calculated from the Stokes–Einstein relation is about equal to that measured, indicating that the calcium ions are rigidly bound and have no residual mobility with respect to the protein's surface. A similar conclusion can be drawn from frequency-dependent T_1 measurements for ^{113}Cd^{2+}-loaded parvalbumin, assuming chemical shift anisotropy is the major relaxation mechanism for this nucleus (Drakenberg, unpublished results).

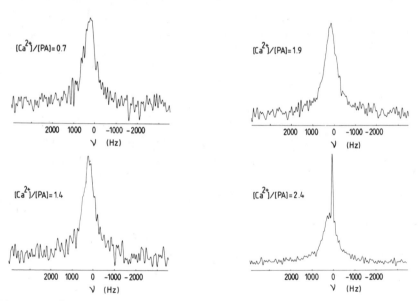

Fig. 7 The ^{43}Ca-NMR spectra of a 5.6 mM parvalbumin solution at 23°C and pH 7.3 at different [Ca^{2+}]/[Parv] ratios as indicated (Reproduced with permission from Andersson et al., 1982a. Copyright 1982 American Chemical Society.)

L. Troponin C

For the discussion here it is sufficient to recall that TnC contains two high-affinity Ca-binding sites ($K_d \approx 10^{-7} M$), which can also bind Mg ions, as well as two additional Ca-binding sites of lower affinity ($K_d \approx 10^{-5} M$) but greater specificity (Potter and Gergely, 1975). Assignment of these binding affinities to the four EF-hand binding sites of the protein was made from studies on proteolytic fragments indicating that the two Ca–Mg sites are sites III and IV, whereas the low-affinity sites are sites I and II (Grabarek et al., 1981; Leavis et al., 1978). The largest conformational changes are caused by the binding of calcium to the two high-affinity sites III and IV, as shown by circular dichroism (CD) techniques (Hincke et al., 1978) and ^{1}H-NMR measurements (Hincke et al., 1981a,b; Levine et al., 1977; Seamon et al., 1977). It has been suggested that only the calcium-specific low-affinity sites are involved in the regulation of muscle contraction because the high-affinity sites might be continuously saturated with Mg^{2+} in the intracellular environment (Potter and Gergely, 1975: Johnson et al. 1981). Suggestions from ^{1}H NMR that binding to Ca^{2+} and Mg^{2+} to the two high-affinity sites causes roughly equal conformational transitions in the apoprotein support such a notion. We shall discuss this point later in more detail.

Detailed ^{23}Na-NMR studies on TnC and two of its tryptic fragments have been reported (Delville et al., 1980a). As indicated earlier in our discussion of the structure of the binding sites, the use of tryptic fragments seems to be valid if only fragments containing two neighboring and related calcium sites are used (see page 165). The fragment TR1 encompasses the low-affinity sites I and II, and the fragment TR2 contains both high-affinity sites III and IV. When using these peptides, one must make the assumption that no large changes have occurred on proteolytic cleavage and that the two halves of the molecule normally react rather independently. Both native TnC and its two fragments interact with sodium ions. The low-affinity sites (I and II) of the fragment TR1 and intact troponin C display competitive Na^+–Ca^{2+} binding. More complicated is the situation for the two high-affinity sites (III and IV) on peptide TR2 and intact troponin. At higher Ca^{2+} concentrations competition between Na^+ and Ca^{2+} for the binding sites is readily apparent, and at low Ca^{2+} concentrations no competition occurs, an observation that is not easily explained (Delville et al. 1980a). These studies demonstrate the usefulness of the study of proteolytic fragments of multisite binding proteins.

Extensive studies with ^{43}Ca NMR have been performed in our laboratory. Studies with slow exchanging calcium ions bound to the high-affinity sites of proteins revealed a R_1 of 1.0×10^3 s^{-1}, which when combined with R_2 deduced from the line width led to an estimate of τ_c of 12 ns, a value expected for a calcium ion tightly bound to and rigidly held by a protein of this size. A similar value was also obtained in a ^{23}Na-NMR study (Delville *et al.*, 1980a). The value obtained for the calcium quadrupole coupling constant (1.05 MHz) is a measure of the symmetry at the binding site (Andersson *et al.*, 1982a). It is of interest that similar values for the EF-hand calcium-binding loops in parvalbumin and calmodulin were observed, whereas the value for prophospholipase A_2, a protein with a different type of binding site, was somewhat lower, indicating greater symmetry of this site. Although it is still too early to draw far-reaching conclusions, these results suggest that non-EF-hand sites may provide a more symmetric surrounding for the calcium ion.

At Ca^{2+}/TnC ratios greater than 2, the line width decreases because of chemical exchange. At fairly high Ca^{2+}/protein ratios the line width is dominated by the exchange between the calcium ion and the two low-affinity sites. From the temperature dependence (Fig. 8), an off-rate for calcium at the two low-affinity sites of 0.6×10^3 s^{-1} was calculated, suggesting an on-rate for calcium on the order of the diffusion limit of the ion in water. The situation with respect to the off-rates was more complicated when Mg^{2+} was present as well, and a different model with markedly different exchange rates had to be invoked in order to fit the data. The pH dependence suggested pK_a values of 4.6 and 6 (Andersson *et al.*, 1981b).

Mg^{2+} and Ca^{2+} compete mostly for the two strong sites III and IV, as competition with Ca^{2+} in ^{25}Mg-NMR experiments indicated. Thus it

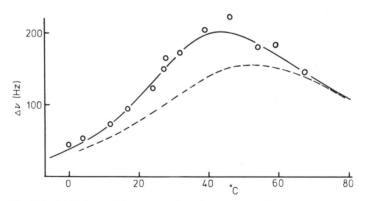

Fig. 8   The ^{43}Ca-NMR line width as a function of temperature for a solution containing 0.9 mM TnC and 6 mM Ca^{2+}. The solid line represents the theoretical curve calculated with the assumption that two sites are exchanging too slowly to affect the band shape and two sites are exchanging at an intermediate rate, $k = 10^3$ s^{-1} at 25°C.

seems likely that broadening of the Mg^{2+} line is caused mainly by the exchange of free Mg^{2+} with Mg^{2+} bound to the high-affinity sites. These two sites seem to be characterized by a pK_a value of ~5.4. The Mg^{2+} off-rate at these sites is ~8×10^3 s^{-1}. This number is not without significance. In fast skeletal muscle, contraction and relaxation occur on a time scale of about 50 ms (McCubbin and Kay, 1980). Several processes have to occur on a short time scale to ensure contraction. Calcium ions must be released, diffuse to the site, and bind to the site; then, if magnesium ions happen to be bound there, they must be released first, and the protein must undergo conformational changes. The dynamics of the conformational changes have been measured by stopped-flow fluorimetry as reported by Johnson et al. (1979). These authors concluded that the rates of release of Mg^{2+} from the high-affinity sites were incompatible with the time scale appropriate to the role of these sites in muscle contraction. Thus Ca^{2+} binding to the low-affinity sites was considered the sole regulatory step in muscle contraction. Binding of Mg^{2+} and Ca^{2+} to the high-affinity sites was suggested to play a structural role only. However, the off-rate we obtained for Mg^{2+} release from the high-affinity sites is not incompatible with a regulatory role of the high-affinity sites in muscle contraction. It is of further interest that the Ca^{2+} exchange rate we measured for the two low-affinity sites, compared to the conformational data of Johnson et al. (1979), suggests that calcium removal from this site may be faster than the conformational change.

^{113}Cd NMR of a mixture of apo-TnC and 4 equiv of Cd^{2+} ions revealed only two resonances in the spectra. The spectrum was identical to that obtained after the addition of only 2 equiv. Thus apparently only Cd^{2+} bound to the two high-affinity sites is in slow exchange and gives rise to two resolved resonances with T_1 values of about 1.7 s each. The failure to detect an observable intensity for the two low-affinity sites of this molecule suggests that the intermediate exchange regime may apply for the free Cd^{2+} and Cd^{2+} bound to the low-affinity sites. Because the chemical shift differences between the free and bound states are rather large (100 ppm), the signal is likely to be too broad to be detectable (Forsén et al., 1979). When the experiment is repeated at lower temperatures, two additional signals of smaller intensity can be detected at a Cd^{2+}/protein ratio of 4. The simplest explanation for this observation is that the exchange between the free Cd^{2+} and the Cd^{2+} bound to the low-affinity sites is sufficiently low to allow detection of all four protein-bound residues (Fig. 9) (Ellis, 1983; Teleman et al., unpublished observation). Calcium is rather effective in competing with Cd^{2+} ions for the strong binding sites: 70% of both Cd resonances disappeared after the addition of 2 equiv of calcium (Forsén et al., 1979). Recent experiments in our laboratory per-

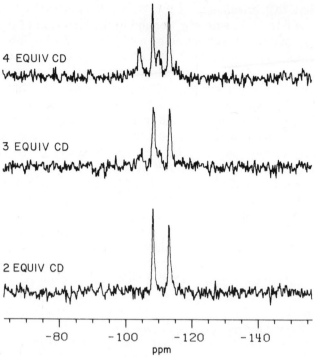

Fig. 9 ^{113}Cd-NMR spectra of TnC at 7°C and pH 6.8. The sample contained 1.7 mM TnC and [Cd]/[TnC] ratios as indicated.

formed with TnC purified from cardiac muscle also showed two clearly resolved ^{113}Cd resonances for this protein. Cardiac TnC binds only three calcium ions, two with higher affinity, since amino acid substitutions have rendered site I inactive (McCubbin and Kay, 1980). Thus our ^{113}Cd-NMR data provide independent support for the idea that sites III and IV of skeletal and cardiac TnC are the two high-affinity sites in both proteins, as earlier concluded from studies with proteolytic fragments (Teleman *et al.*, 1983).

M. *Calmodulin*   (See note at end of chapter)

Calmodulin can substitute for TnC in some functions in the control of muscle relaxation, indicating the high degree of homology between these two proteins (Amphlett *et al.*, 1976). Indeed, both proteins have very similar molecular weights (MW 17,000) and, just as in the case of TnC, four EF-hand structures can be observed in the amino acid sequence of calmodulin. However, in contrast to TnC, these four calcium-binding sites are not easily differentiated into high- and low-affinity sites. The exact values of the binding constants are still a debated issue, and the results appear to be very sensitive to the nature of the buffer and to the presence of other cations such as K^+ and Mg^{2+}. In the case of TnC, biochemical studies with the proteolytic fragments TR1 and TR2 were decisive in assigning the binding affinities to the binding sites (see previous paragraph), but this approach was less successful here, because all four sites appear to have very similar binding constants (Drabikowski *et al.*, 1977; Walsh *et al.*, 1977). Physical studies, most notably those utilizing the fluorescent properties of the lanthanide ion terbium as a replacement for calcium ions, indicate that sites I and II could be filled before sites III and IV (see Haiech *et al.*, 1981; Kilhoffer *et al.*, 1980; Wallace *et al.*, 1982; Wang *et al.*, 1982). It should be noted that this would be the reverse situation with respect to TnC, where sites III and IV are the high-affinity sites.

On binding of Ca^{2+}, calmodulin undergoes conformational changes that can be followed by several spectroscopic techniques, including ^{1}H-NMR spectroscopy (Hincke *et al.*, 1981a; Ikura *et al.*, 1983; Krebs and Carafoli, 1982; Seamon, 1980), by changes in the exposure of surface amino acids as measured by chemical modification techniques (Klee *et al.*, 1980; Walsh *et al.*, 1978), or by laser photochemically induced dynamic nuclear polarization experiments (Hincke *et al.*, 1981b). It is generally agreed that a hydrophobic surface is exposed on binding of calcium. This surface appears to be involved in the interaction between calmodulin and its target proteins (LaPorte *et al.*, 1980; Marshak *et al.*, 1981). It has been suggested that the Met-71, -72, and -76 constitute part of this area (Walsh *et al.*, 1978). A large variety of hydrophobic molecules, such as antipsychotic drugs of the phenothiazine type (for review see Klee *et al.*, 1980), the antihypertensive drug Felodipine (Bolström *et al.*, 1981), various fluorescent dyes and calcium antagonists (Andersson *et al.*, 1983a), and the hormone β-endorphin (Sellinger-Barnette and Weiss, 1982), can bind to calmodulin and block its interaction with certain target enzymes (see Klee *et al.*, 1980). All these agents have a much lower affinity for TnC, suggesting that this protein does not have a similar hydrophobic surface and providing an explanation of why TnC usually cannot substitute for calmodulin.

^{23}Na-NMR measurements of cation binding to calmodulin and two of its proteolytic fragments have been reported (Delville *et al.*, 1980b). Proteolytic fragment 1 encompasses sites I and II, whereas fragment 2 con-

tains both sites III and IV. The relaxation rate enhancement of the ^{23}Na nuclei, as measured from the line broadening of the observed resonance on the addition of protein, shows that Na$^+$ binds to calmodulin and its two fragments. Ca^{2+} and Mg^{2+} compete with Na$^+$ for the same binding sites. The affinity of Ca^{2+} for binding sites I and II is about 10^{-6} M, and for binding sites III and IV about 10^{-5} M. Thus these data indicate that sites I and II of calmodulin are filled by calcium ions before sites III and IV. The observed competition among Na$^+$, Mg^{2+}, and Ca^{2+} for the same binding sites allows a fine-tuning of the physiological response on release of Ca^{2+} (Delville et al., 1980b). It would be of interest to use the same ^{23}Na-NMR competition technique (and possibly the very insensitive ^{39}K-NMR) to study the interaction of potassium ions with calmodulin. Direct binding and competition studies have suggested that the protein has a rather high affinity for this univalent cation (Haiech et al., 1981).

^{43}Ca-NMR studies on calcium binding to calmodulin purified from bovine brain or testicles have been reported (Andersson et al., 1982a, 1982c). A study on the slow exchanging calcium ions bound to the two sites with the highest affinity resulted in $R_1 = 1.2 \times 10^3$ s^{-1}. Combined with the line width of 770 Hz, which corresponds to $R_2 = 2.4 \times 10^3$ s^{-1}, this figure indicates a correlation time for the bound calcium ions of 8 ns which is a value approximating that expected for the overall rotational correlation time of the protein (Andersson et al., 1982a) and close to that obtained from the ^{23}Na-NMR study (Delville et al., 1980b).

As with parvalbumin and TnC, the resonance of the bound calcium ion is shifted 10 ppm downfield from the signal of free Ca^{2+}. Unfortunately this value is not easily interpreted compared with chemical shifts obtained for small calcium complexes (see Chapter 8, Vol. II on calcium and magnesium NMR). At higher Ca^{2+}/calmodulin ratios, the line shifts to the position of the free Ca^{2+} ion and decreases in line width. This effect can be attributed to chemical exchange effects. Figure 10 shows a titration of ^{43}Ca into a solution of calmodulin. The observed line width remains at a plateau until a ratio of about 4 calcium ions per calmodulin molecule. This observation does not mean that all four sites are in slow exchange. In fact, analysis of the shape of the binding curve, as well as of the temperature dependence of the line width, suggests that calmodulin has, like TnC, two high-affinity sites that are in the slow-exchange regime and two low-affinity sites that are in the fast-to-intermediate exchange regime (Andersson et al., 1982c; Drakenberg et al., 1983). We shall consider this in more detail in the following discussion of the ^{113}Cd-NMR results.

Competition experiments with Mg^{2+} indicate that affinity for this metal ion is less than that for Ca^{2+}. The addition of Zn^{2+} gave rise to results that

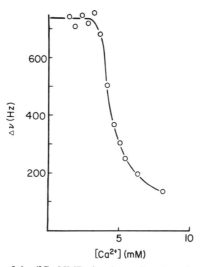

Fig. 10 The line width of the ^{43}Ca-NMR signal as a function of calcium concentration for a solution containing 0.86 mM calmodulin and 17.5 mM Mg^{2+} at pH 7.0 and t = 23°C. The solid curve is calculated using complete band shapes and assuming four binding sites per protein. Two metal ions exchange with k_{off} = 1.1 × 10^3 s^{-1} and two with k_{off} ≈ 10 s^{-1}. χ = 1.03 MHz and K_a > 10^4 M^{-1}.

are not easily interpreted (for discussion see Andersson *et al.*, 1982c). The pK_a value determined for the two low-affinity sites is 4.4. Addition of the antipsychotic drug trifluoperazine markedly reduces the excess line width measured from the calcium spectra (Thulin *et al.*, 1980). Since this broadening is mainly caused by the exchange of free Ca^{2+} ions with Ca^{2+} bound to the two weaker sites, this indicates that the binding of trifluoperazine reduces this exchange rate so that all four sites are in slow exchange, as reported earlier for ^{113}Cd-NMR spectra (Forsén *et al.*, 1980; Sudmeier *et al.*, 1980).

Shimizu *et al.* (1982) studied calmodulin purified from *Tetrahymena*. Calmodulin has been remarkably conserved during evolution. The largest differences are between the proteins purified from bovine brain and *Tetrahymena:* Only 12 amino acids have been altered by evolutionary pressure. The ^{43}Ca-NMR results obtained with *Tetrahymena* calmodulin (unfortunately studied only at a high Ca^{2+}/protein ratio) are quite similar to those described earlier for the bovine preparations. It should be stated that these authors interpret the observed increased line width as indicating an exchange between both the high- and low-affinity sites, whereas our earlier results demonstrated that only the latter two contributed significantly.

Another difference is that these authors interpret the decrease in line width after the addition of trifluoperazine to the liberation of Ca^{2+}, an effect not easily reconciled with the available ^{113}Cd-NMR data.

^{113}Cd-NMR spectra have been reported for calmodulin purified from bovine brain and testicles (Forsén et al., 1980; Sudmeier et al., 1980; Andersson et al., 1982c). For these preparations only two well-resolved resonances for protein-bound residues are observed on the addition of 2 or 4 equiv or more of cadmium. These resonances represent the Cd^{2+} bound to the two calmodulin sites with the highest affinity. The resonances for the two sites with lower affinity are, as for TnC, broadened by exchange and escape detection at higher frequency but can be visualized as very broad resonances at lower fields (Forsén et al., 1980; Sudmeier et al., 1980). Unlike the observations on TnC, no additional resolved resonances were seen for the low-affinity sites at lower temperatures. It should also be mentioned that, although most biochemical experiments indicate rather similar behavior for all four binding sites of calmodulin, the available ^{43}Ca, and ^{113}Cd data suggest that, as with TnC, the sites can be divided into two high-affinity (slow-exchange) and two low-affinity (fast-to-intermediate-exchange) sites. The NMR method may be more sensitive to subtle differences in calcium affinity (as measured from the exchange) than the normally applied biochemical techniques for measurements of binding constants.

Interesting results were obtained after the addition of hydrophobic agents such as the antipsychotic drugs trifluoperazine and chlorpromazine. On binding of two molecules of this agent, the exchange rates for Cd^{2+} bound to the two fast exchanging sites dropped significantly, because four resolved signals then appeared in the spectra (Forsén et al., 1980; Sudmeier et al., 1980). Similar changes were observed after the addition of a hydrophobic fluorescent dye or a calcium antagonist (D600). These were paralleled by characteristic changes in the high-resolution ^1H-NMR spectra of the protein (Andersson et al., 1983a). In contrast, binding of the experimental antihypertensive drug Felodipine resulted in a ^{113}Cd-NMR spectrum that did not resemble the one obtained after the addition of trifluoperazine (Boström et al., 1981). Thus calmodulin is capable of adopting different conformations on binding of different agents. Another set of ^{113}Cd-NMR experiments were concerned with the replacement of protein-bound ^{113}Cd with lanthanide ions. Both tightly bound Cd^{2+} ions are replaced preferentially and simultaneously by lanthanide ions. All lanthanides substitute in an identical fashion, in contrast to our results with parvalbumin. Thus replacement of Ca^{2+} by Tb^{3+}, as used by many researchers, appears to be a valid approach in the study of calmodulin (Kilhoffer et al., 1980; Wallace et al., 1982; Wang et al., 1982).

V. Future Studies

The discussion of ^{23}Na-, ^{25}Mg-, ^{43}Ca-, and ^{113}Cd-NMR studies on calcium-binding proteins in Section IV should suffice to underline the usefulness of such studies. All ^{113}Cd-NMR shifts measured thus far agree with the notion that calcium ligands are solely oxygen atoms. The proteins prothrombin (Marsh *et al.*, 1979) and S100 (Isobe and Okuyama, 1978, 1981) are the only ones having ligands with a pK_a of 7.0 or higher implicated in the binding of such ions. Thus it would be of great interest to obtain ^{113}Cd-NMR chemical shifts for these proteins. Further use of ^{113}Cd-NMR lies in the study of peptides. If peptide TR2 from TnC has the same binding characteristics as the native protein, it should give rise to well-resolved resonances, whereas troponin's TR1 peptide would not be expected to show any resolved resonances. The feasibility of such studies with peptide fragments has already been demonstrated by other NMR studies (Delville *et al.*, 1980a, b, Andersson *et al.*, 1983b). They should be extended with ^{43}Ca- and ^{25}Mg-NMR studies such as those described in Section IV. However, in such peptide fragments it will be less problematic to identify the sites mainly contributing to the exchange than in the native protein. Thus such studies provide a necessary check on the interpretation of earlier experiments.

In contrast to studies on small proteolytic fragments, in the future attention should be focused on proteins of higher molecular weight. This will probably not cause dramatic problems in the study of the ^{113}Cd nucleus, but studies on ^{43}Ca and ^{25}Mg are close to the presently available reliable detection limits. Improvements in instrumentation must lead to probe designs that allow shorter dead times for low resonance frequencies, so that resonances with a line width larger than the current practical detection limit of 1000 Hz can be reliably detected and characterized. Such developments are of the utmost importance for future ^{43}Ca- and ^{25}Mg-NMR experiments and may eventually possibly allow the study of other protein-bound metal nuclei (e.g., ^{25}Mg^{2+}, ^{139}Ln^{3+}, ^{67}Zn^{2+}).

Not only will larger proteins need to be investigated in the near future, but many biochemically interesting protein complexes will have to be studied. For example, the ubiquitously distributed contractile protein actin is reported to form a tight 1 : 1 complex with bovine pancreatic DNase I, both proteins containing tightly bound calcium ions (Liao *et al.*, 1982; Suck *et al.*, 1981). Of course, of the utmost importance are studies with the modulator proteins TnC and calmodulin. ^{113}Cd- and ^{43}Ca-NMR measurements of intact troponin (i.e., the 1 : 1 : 1 complex of TnC, TnI, and TnT) are important, because it has been reported that the calcium-binding affinity of TnC is increased in the complex. The complexes between calmodulin and histone 2B or the myelin basic protein should be amenable to study (Grand and Perry, 1980). More demanding but probably of greater physiological significance is the study of the ternary complex of calcium–calmodulin–myosin light-chain kinase (Blumenthal and Stull, 1982). In some cases the solubility of such complexes is poor and will prohibit studies. In such cases the application of solid state NMR techniques may be required.

As indicated in Section I, total intracellular Mg^{2+} concentrations are 1–5 mM, whereas calcium levels never exceed 10^{-5} M. Thus a careful assessment of binding constants, on-rates, and off-rates, which as previously outlined can be determined only from direct metal NMR techniques, is required to understand the role of the various calcium-binding sites of a multi-calcium-site protein *in vivo*. For example, parvalbumin binds calcium ions so tightly that it should be capable of preferably binding calcium ions in muscle cells over TnC. It has been proposed that parvalbumin is responsible for relaxation in fast muscle cells. In such a role, the Ca^{2+} ions released from the sarcoplasmic reticulum must first bind to the TnC, causing the conformational changes responsible for contraction, before parvalbumin can bind the Ca^{2+} ions. Estimates of the binding constants and metal on- and off-rates, as well as the rates of protein conformational changes (parvalbumin, Birdsall *et al.*, 1979; TnC, Johnson *et al.*, 1979), will allow simulation of the sequence of events in muscle contraction and relaxation and a comparison with the time scales observed in physiological experiments (50 ms). This would be of help in differentiating between a role for parvalbumin in muscle relaxation only and a role as a general calcium diffusion agent.

Since no isotopes with spectroscopic properties are known for calcium or magnesium, direct investigation of the nuclei will be possible only by studying the two available isotopes with a quadrupole moment, ^{43}Ca and ^{25}Mg. Enrichment has been and will be necessary for all studies of ^{43}Ca because its natural abundance is only 0.14%. ^{25}Mg enrichment is normally helpful, because its natural abundance is about 10%. Although other nuclei with more favorable properties (like ^{113}Cd or lanthanides) may be used as substitutes, one will always have to confirm that their deployment is allowed. Our results with lanthanides binding to parvalbumin illustrate the possible problems one may encounter. Finally, we can state with confidence that NMR study of less receptive nuclei will continue to provide valuable insights into the complex role of calcium's regulatory functions in biological processes.

References

Adelstein, R. S., and Eisenberg, E. (1980). *Ann. Rev. Biochem.* **49**, 921–956.
Amphlett, G. W., Vanaman, T. C., and Perry, S. V. (1976). *FEBS Lett.* **72**, 163–158.

Andersson, A., Drakenberg, T., Forsén, S., and Thulin, E. (1983a) *Eur. J. Biochem.* In press.

Andersson, A., Forsén, S., Thulin, E., and Vogel, H. J. (1983b). *Biochemistry* **22**, 2309–2313.

Andersson, T. (1981). Ph.D. thesis, Univ. of Lund, Lund, Sweden.

Andersson, T., Drakenberg, T., Forsén, S., Wieloch, T., and Lindström, M. (1981a). *FEBS Lett.* **123**, 115–117.

Andersson, A., Drakenberg, T., Forsén, S., and Thulin, E. (1981b). *FEBS Lett.* **125**, 39–43.

Andersson, T., Drakenberg, T., Forsén, S., Thulin, E., and Swärd, M. (1982a). *J. Am. Chem. Soc.* **104**, 576–580.

Andersson, T., Chiancone, E., and Forsén, S. (1982b). *Eur. J. Biochem.* **125**, 103–108.

Andersson, T., Drakenberg, T., Forsén, S., and Thulin, E. (1982c). *Eur. J. Biochem.* **126**, 501–505.

Antonini, E., and Chiancone, E. (1977). *Ann. Rev. Biophys. Bioeng.* **6**, 239–271.

Bailey, D. B., Ellis, P. D., Cardin, A. D., and Behnke, W. D. (1978). *J. Am. Chem. Soc.* **100**, 5236–5237.

Becker, J. W., Reeke, G. N., Wang, J. L., Cunningham, D. A., and Edelman, G. M. (1975). *J. Biol. Chem.* **250**, 1513–1520.

Bennick, A., McLaughlin, A. C., Grey, A. A., and Madapallimattam, G. (1981). *J. Biol. Chem.* **256**, 4741–4746.

Birdsall, W. J., Levine, B. A., Williams, R. J. P., Demaille, J. G., Haiech, J., and Pechere, J. F. (1979). *Biochimie* **61**, 741–752.

Bittiger, H., ed. (1977). "Concanavalin A as a Tool." Wiley, New York.

Blumenthal, D. K., and Stull, Y. T. (1982). *Biochemistry* **21**, 2386–2392.

Blundell, T., Dodson, G., Hodgkin, D., and Mercola, D. (1972). *Adv. Protein Chem.* **26**, 279–301.

Boström, S. L., Ljung, B., Mårdh, S., Forsén, S., and Thulin, E. (1981). *Nature* **292**, 777–779.

Cavé, A., Parello, J., Drakenberg, T., Thulin, E., and Lindman, B. (1979). *FEBS Lett.* **100**, 148–152.

Cavé, A., Saint-Yves, A., Parello, J., Swärd, M., Thulin, E., and Lindman, B. (1982). *Mol. Cell. Biochem.* **44**, 161–172.

Celio, M. R., and Heizmann, C. W. (1982). *Nature* **297**, 504–506.

Cheung, W. Y. (1980). *Science* **207**, 19–27.

Cheung, W. Y. (1982). *Sci. Am.* **6**, 48–56.

Civan, M. M., and Shporer, M. (1978). *In* "Biological Magnetic Resonance" (J. Berliner and J. Reuben, eds.), Vol. 1, pp. 1–32. Plenum, New York.

Clarke, M., and Spudich, J. A. (1977). *Ann. Rev. Biochem.* **46**, 797–822.

Cliffe, S. G. R., and Grant, D. A. W. (1981). *Biochem. J.* **193**, 655–658.

Coffee, C. J., and Solano, C. (1976). *Biochem. Biophys. Acta* **453**, 67–75.

Cohen, P. (1980). *Eur. J. Biochem.* **111**, 563–571.

Cohn, M., and Hughes, T. R. (1962). *J. Biol. Chem.* **237**, 176–181.

Cookson, D. J., Levine, B. A., Williams, R. J. P., Jontell, M., Linde, A., and de Bernard, B. (1980). *Eur. J. Biochem.* **110**, 273–278.

Cotton, F. A., and Hasen, E. E. (1971). *In* "The Enzymes" (P. D. Boyer, ed.), Vol. IV, 3rd Edition, pp. 153–176. Academic Press, New York.

Delville, A., Grandjean, J., Laszlo, P., Gerday, C., Grabarek, Z., and Drabikowski, W. (1980a). *Eur. J. Biochem.* **105**, 289–295.

Delville, A., Granjean, J., Laszlo, P., Gerday, C., Breszka, H., and Drabikowski, W. (1980b). *Eur. J. Biochem.* **109**, 515–522.

Derancourt, J., Haiech, J., and Pechère, J. F. (1978). *Biochim. Biophys.* **532**, 373–380.
Dijkstra, B. W., Kalk, K. H., Hol, W. G. J., and Drenth, J. (1981). *J. Mol. Biol.* **147**, 97–123.
Drabikowski, W., Kuznicki, J., and Grabarek, Z. (1977). *Biochim. Biophys. Acta* **485**, 124–133.
Drakenberg, T. (1982). *Acta Chem. Scand.* **A36**, 79–82.
Drakenberg, T. and Vogel, H. J. (1983). *In* "Calcium binding proteins in Health and Disease." (Siegel, F. L. *et al.* eds.), Elsevier, Amsterdam. In press.
Drakenberg, T., Lindman, B., Cavé, A., and Parello, J. (1978). *FEBS Lett.* **92**, 346–350.
Drakenberg, T., Forsén, S., and Lilja, H. (1983). *J. Magn. Reson.* to be published.
Ellis, P. D. (1983). *In* "The Multinuclear Approach to NMR Spectroscopy" (J. B. Lambert and F. G. Siegel, eds.) pp. 457–523. D. Reidel Publ. Co., Boston.
Epstein, M., Reuben, J., and Levitzki, A. (1977). *Biochemistry* **16**, 2449–2457.
Forsén, A., and Lindman, B. (1981). "Method of Biochemistry Analysis" (D. Glick, ed.), Vol. 27, pp. 289–486. Wiley, New York.
Forsén, S., Thulin, E., and Lilja, H. (1979). *FEBS Lett.* **104**, 123–126.
Forsén, S., Thulin, E., Drakenberg, T., Krebs, J., and Seamon, K. (1980). *FEBS Lett.* **117**, 184–194.
Forsén, S., Andersson, T., Drakenberg, T., Thulin, E., and Wieloch, T. (1981). *In* "Advances in Solution Chemistry" (I. Dentini *et al.*, eds.), pp. 191–208. Plenum, New York.
Furie, B. C., Blumenstein, M., and Furie, B. (1979). *J. Biol. Chem.* **254**, 12521–12530.
Gadian, D. G., and Radda, G. K. (1981). *Ann. Rev. Biochem.* **50**, 69–84.
Gariépy, J., Sykes, B. D., Reid, R. E., and Hodges, R. S. (1982). *Biochemistry* **21**, 1506–1512.
Gerday, C., and Gillis, J. M. (1976). *J. Physiol.* **258**, 96–97.
Goldie, A. H., and Sanwal, B. D. (1980). *J. Biol. Chem.* **255**, 1399–1405.
Grabarek, Z., Drabikowski, W., Virokurov, L., and Lu, R. C. (1981). *Biochim. Biophys. Acta* **671**, 227–233.
Grand, R. J. A., and Perry, S. V. (1980). *Biochem. J.* **189**, 227–240.
Grandjean, J., Laszlo, P., and Gerday, C. (1977). *FEBS Lett.* **81**, 376–380.
Grizutti, K., and Perlmann, G. E. (1975). *Biochemistry* **14**, 2171–2175.
Gupta, R. J., Benovic, J. L., and Rose, Z. B. (1978). *J. Biol. Chem.* **253**, 6165–6171.
Haiech, J., Derancourt, J., Pechère, J. F., and Demaille, J. G. (1979). *Biochemistry* **18**, 2752–2758.
Haiech, J., Klee, C. B., and Demaille, J. G. (1981). *Biochemistry* **20**, 3890–3897.
Hauschka, P. V., and Carr, S. A. (1982). *Biochemistry* **21**, 2538–2547.
Hauschka, P. V., and Gallop, P. M. (1977). In "Calcium-binding Proteins and Calcium Function" (R. H. Wasserman *et al.*, eds.), pp. 338–347. Elsevier, Amsterdam.
Hincke, M. T., McCubbin, W. D., and Kay, C. M. (1978). *Can. J. Biochem.* **86**, 384–395.
Hincke, M. T., Sykes, B. D., and Kay, C. M. (1981a). *Biochemistry* **20**, 3286–3294.
Hincke, M. T., Sykes, B. D., and Kay, C. M. (1981b). *Biochemistry* **20**, 4185–4193.
Hol, W. G. J., Van Duijnen, P. T., and Berendsen, H. J. C. (1978). *Nature* **273**, 443–446.
Horrocks, W., and Sudnick, D. R. (1981). *Acc. Chem. Res.* **14**, 384–392.
Ikura, M., Hiraoki, T., Hikichi, K., Mikuni, T., Yazawa, M., and Yagi, K. (1983). *Biochemistry.* **22**, 2573–2579.
Isobe, T., and Okuyama, T. (1978). *Eur. J. Biochem.* **89**, 379–380.
Isobe, T., and Okuyama, T. (1981). *Eur. J. Biochem.* **116**, 79–86.
Johnson, J. D., Charlton, S. C., and Potter, J. D. (1979). *J. Biol. Chem.* **254**, 3479–3502.
Johnson, J. D., Collins, J. H., Robertson, S. P. and Potter, J. D. (1981). *J. Biol. Chem.* **255**, 9635–9640.
Kalk, A., and Berendsen, H. J. C. (1976). *J. Magn. Reson.* **24**, 343–366.

Kilhotter, M.-C., Gerard, D. and Demaille, J. G. (1980). *FEBS Lett.* **120**, 99–103.

Klee, C. A., Crouch, T. H., and Richman, P. G. (1980). *Ann. Rev. Biochem.* **49**, 489–515.

Krebs, E. G., and Beavo, J. A. (1979). *Ann. Rev. Biochem.* **48**, 923–961.

Krebs, J., and Carafoli, E. (1982). *Eur. J. Biochem.* **124**, 619–627.

Kretsinger, R. H. (1976). *Ann. Rev. Biochem.* **45**, 239–266.

Kretsinger, R. H. (1980). *CRC Crit. Rev. Biochem.* 114–174.

Kretsinger, R. H., and Barry, C. D. (1975). *Biochim. Biophys. Acta* **405**, 40–52.

Kretsinger, R. H., and Nockolds, C. E. (1973). *J. Biol. Chem.* **248**, 3312–3314.

LaPorte, D. C., Wiesman, B. M., and Storm, D. R. (1980). *Biochemistry* **19**, 3814–3819.

Leavis, P. C., Rosenfeld, S. S., Gergely, J., Grabarek, Z., and Drabikowski, W. (1978). *J. Biol. Chem.* **253**, 5452–5459.

Lee, L., and Sykes, B. D. (1981). *Biochemistry* **20**, 1156–1162.

Levine, B. A., Mercola, D., Coffman, D., and Thornton, J. M. (1977). *J. Mol. Biol.* **115**, 743–760.

Liao, T. H., Ting, R. S., and Yeung, J. E. (1982). *J. Biol. Chem.* **257**, 5637–5644.

Linde, A., Bhown, M., and Butler, W. T. (1980). *J. Biol. Chem.* **255**, 5931–5942.

McCubbin, W. D., and Kay, C. M. (1980). *Acc. Chem. Res.* **13**, 185–192.

McLennan, D. H., and Holland, P. C. (1975). *Ann. Rev. Biophys. Bioeng.* **4**, 377–404.

Marsh, H. C., Robertson, P., Scott, M. E., Koehler, K. A., and Hiskey, R. G. (1979). *J. Biol. Chem.* **254**, 10260–10275.

Marshak, D. R., Watterson, D. M., and van Eldik, L. J. (1981). *Proc. Nat. Acad. Sci. USA* **78**, 6793–6797.

Martin, R. B., and Richardson, F. S. (1979). *Q. Rev. Biophys.* **2**, 181–209.

Matthews, B. W., and Weaver, L. H. (1974). *Biochemistry* **13**, 1719–1725.

Maximov, E. E., and Mitin, Y. V. (1979). *Biochimie* **61**, 751–754.

Mildvan, A. S. (1979). *Adv. Enzymol.* **49**, 103–126.

Nelson, D. J., Opella, S. J., and Jardetzky, O. (1976). *Biochemistry* **15**, 5552–5560.

Norne, J. E., Gustavsson, H., Forsén, S., Chiancone, E., Kuiper, H. A., and Antonini, E. (1979). *Eur. J. Biochem.* **98**, 591–595.

Palmer, A. R., Bailey, D. B., Behnke, W. D., Cardin, A. D., Yang, P. P., and Ellis, P. D. (1980). *Biochemistry* **19**, 5063–5070.

Parello, J., Reimarsson, P., Thulin, E. and Lindman, B. (1979). *FEBS Lett.* **100**, 152–156.

Pechère, J. F. (1977). *In* "Calcium-binding Proteins and Calcium Function" (R. H. Wasserman *et al.*, eds.), pp. 213–221. Elsevier, Amsterdam.

Perry, S. V. (1979). *Biochem. Soc. Trans.* **7**, 594–616.

Pletcher, C. H., Bouthouses-Brown, E. F., Bryant, R. G., and Nelsestuen, G. L. (1981). *Biochemistry* **20**, 6149–6155.

Potter, J. D., and Gergely, J. (1975). *J. Biol. Chem.* **250**, 4628–4633.

Ragg, E., Cavé, A., Drakenberg, T. (1984). In preparation.

Reid, R. E., Gariépy, J., Saund, A. K., and Hodges, R. S. (1981). *J. Biol. Chem.* **256**, 2742–2751.

Robertson, P., Hiskey, R. G., and Koehler, K. L. (1978). *J. Biol. Chem.* **253**, 5880–5883.

Robertson, P., Koehler, K. A., and Hiskey, R. G. (1979). *Biochem. Biophys. Res. Commun.* **86**, 265–270.

Seamon, K. (1980). *Biochemistry* **19**, 207–215.

Seamon, K., Hartshorne, D. J., and Bothner-By, A. A. (1977). *Biochemistry* **16**, 8691–8697.

Sellinger-Barnette, M., and Weiss, B. (1982). *Mol. Pharm.* **21**, 86–91.

Shimizu, T., Hatano, M., Nagao, S., and Mozawa, Y. (1982). *Biochem. Biophys. Res. Comm.* **106**, 1112–1118.

Slotboom, A. J., Jansen, E. H. J. M., Vlijm, M., Pattens, F., Soares de Aranjo, P., and de Haas, G. H. (1978). *Biochemistry* **17**, 4593–4600.

Sowadski, J., Cornick, G., and Kretsinger, R. H. (1978). *J. Mol. Biol.* **124,** 123–132.
Stenflo, J., and Suttie, J. W. (1977). *Ann. Rev. Biochem.* **46,** 157–172.
Stewart, A. A., Ingebritsen, T. S., Manalan, A., Klee, C. B., and Cohen, P. (1982). *FEBS Lett.* **137,** 80–84.
Suck, D., Kabsch, W., and Mannherz, G. (1981). *Proc. Nat. Acad. Sci. USA* **78,** 4319–4323.
Sudmeier, J. L., Evelhoch, J. L., Bell, S. J., Storm, M. C., and Dunn, M. F. (1980). *In* "Calcium-binding Proteins: Structure and Function" (F. M. Siegel *et al.,* eds.), pp. 235–237. Elsevier, Amsterdam.
Sudmeier, J. L., Bell, S. J., Storm, M. C., and Dunn, M. F. (1981). *Science* **212,** 560–562.
Sykes, B. D., Hull, W. E., and Snyder, G. H. (1978). *Biophys. J.* **21,** 137–146.
Szebenyi, D. M. E., Obendorf, S. K., and Moffat, K. (1981). *Nature* **294,** 327–332.
Teleman, O., Drakenberg, T., Forsén, S., and Thulin, E. (1983). *Eur. J. Biochem.* In press.
Thulin, E., Forsén, S., Drakenberg, T., Andersson, T., Seamon, K. B., and Krebs, J. (1980). *In* "Calcium Binding Proteins: Structure and Function" (F. M. Siegel *et al.,* eds.), pp. 243–244. Elsevier, New York.
Tufty, R. M., and Kretsinger, R. H. (1975). *Science* **187,** 166–169.
Verheij, H. M., Volwerk, J. J., Jansen, E. H. J. M., Puijck, W. C., Dijkstra, B. W., Drenth, J., and de Haas, G. H. (1980). *Biochemistry* **19,** 743–750.
Vogel, H. J. (1982). Biochemistry, **22,** 668–674.
Vogel, H. J., and Bridger, W. A. (1982a). *Biochemistry* **21,** 394–401.
Vogel, H. J., and Bridger, W. A. (1982b). *Biochemistry* **21,** 5825–5831.
Vogel, H. J., Lindahl, L., and Thulin, E. (1983). *FEBS Lett.* **157,** 241–246.
Vogt, H. P., Strassburger, W., Wollmer, A., Fleischhauer, J., Bullard, B., and Mercola, D. (1979). *J. Theor. Biol.* **76,** 297–310.
Wallace, R. E., Tallant, E. A., Dockter, M. E., and Cheung, W. Y. (1982). *J. Biol. Chem.* **257,** 1845–1854.
Walsh, M., Stevens, F. C., Kuznicki, J., and Drabikowski, W. (1977). *J. Biol. Chem.* **252,** 7440–7443.
Walsh, M., Stevens, F. C., Oikawa, K., and Kay, C. M. (1978). *Biochemistry* **17,** 3924–3930.
Wang, G. L. A., Aquaron, R. R., Leavis, P. C., and Gergely, J. (1982). *Eur. J. Biochem.* **124,** 7–12.
Wassserman, R. H. (1980). *In* "Calcium-binding Proteins: Structure and Function" (F. L. Siegel *et al.,* eds.) pp. 357–361. Elsevier, Amsterdam.
Waxman, L. (1981). *Methods Enzymol.* **80,** 664–680.
Wu, S. T., Pieper, G. M., Salhany, J. M., and Elliot, R. S. (1981). *Biochemistry* **20,** 7399–7403.
Zanetti, M., de Bernard, B., Jontell, M., and Linde, A. (1981). *Eur. J. Biochem.* **113,** 541–545.

Recent studies in our laboratory using tryptic fragments of calmodulin have provided compelling evidence for sites III and IV being the strong calcium binding sites in agreement with ^{1}H studies (Seamon, 1980; Ikura *et al.,* 1983). The fragments encompassing sites III and IV gave rise to a [113]Cd spectrum identical to that observed for the addition of the first two Cd^{2+} ions to intact calmodulin. Identically to the filling of the weaker sites of the intact protein, no resonance was observed for the amino-terminal half peptide comprising domains I and II (Anderson *et al.,* 1983b). Further studies, showed that each half of calmodulin contains one Ca^{2+}-exposed hydrophobic domain and that binding of trifluoperazine caused similar changes in the [113]Cd NMR spectra of the fragments and the intact protein. The latter units agree nicely with those obtained using affinity-chromatography (Vogel *et al.,* 1983).

8 Amphiphilic and Polyelectrolyte Systems

Björn Lindman

Physical Chemistry 1
University of Lund
Lund, Sweden

I. Polyelectrolytes: Polycharged Macromolecules and Aggregates of Ionic Amphiphiles

There are a multitude of *polycharged systems* that are of interest in various fields of chemistry, biology, and chemical engineering. In a wide range of suspensions the dispersed solid particles acquire a high charge density (e.g., colloidal precipitates, clays, and colloidal silica) and, furthermore, it is a common procedure to create high charge densities (e.g., by surfactant adsorption) to stabilize suspensions. There are also many thermodynamically stable polycharged systems, the most significant ones being polymers with charged repeating units and systems of ionic am-

NMR OF NEWLY ACCESSIBLE NUCLEI, VOL. 1

phiphiles with extensive self-association. Although the term "polyelectrolyte" could thus logically be used to describe a wide range of phenomena, it is common practice to use it only for multicharged polymers. We shall follow this usage in this chapter which describes multinuclear NMR studies on solutions of polycharged macromolecules as well as on systems of ionic amphiphiles. Other polycharged systems, such as thermodynamically unstable suspensions, have been little studied by NMR, but this area is a growing field where the same principles apply as those described here. In our discussion, it will be of interest to consider only aqueous systems, and all the examples given thus refer to aqueous solutions. Nuclear magnetic resonance studies on these systems constitute a very active field, and the literature is growing rapidly. Therefore no attempt has been made to cover all the published work. Instead, a few illustrative studies will be cited, and it should be realized that in several cases other similar work has been performed. More comprehensive coverages of the literature exist and will be referred in the following discussion.

Both *synthetic* and *biological* polyelectrolytes are of interest. Frequently occurring charged groups in the former are, for polyanions, carboxylate [as in polyacrylate and polymethacrylate (PMA)], sulfonate (polystyrene sulfonate), and sulfate (polyvinyl sulfate), and, for polycations, positively charged nitrogen (polyimines and polyamines). The most important classes of biological polyions are polysaccharides and nucleic acids (DNA and RNA). In certain respects, synthetic polyelectrolytes can serve as suitable model systems for biological ones. Relevant reviews providing an introduction to several aspects of polyelectrolytes are available (Eisenberg, 1977; Oosawa, 1971; Overbeek, 1976; Manning, 1978; Manning, 1979). *Amphiphilic compounds* can have a wide range of structures and be of either synthetic or biological origin. In spite of their chemical diversity, different amphiphile–water systems behave similarly in important respects. A *surfactant* often consists of a polar group, such as sulfate or carboxylate, and a linear alkyl chain. Its behavior in aqueous solution is dictated by the entropically unfavorable water–hydrocarbon contact. By extensive self-association, the water–hydrocarbon contact is minimized, whereas the polar group is hydrated. Self-association starts with the formation of spherical *micelles* from a rather well-defined concentration, the critical micelle concentration (cmc), and proceeds at higher concentrations or in the presence of a third component with the formation of various *liquid crystalline phases* and *reversed micelles*. The phase behavior that serves as a basis for all physicochemical studies is exemplified in Fig. 1. *Biological amphiphiles* such as lecithin display a lamellar liquid crystalline phase over a wide composition range. This lamellar phase has the type of lipid bilayer structure that is the building block of most *biological membranes*. The rather nonspecific self-associa-

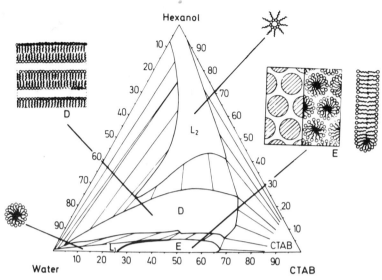

Fig. 1 Phase diagram (at 25°C) from the study of Ekwall and co-workers (1975) on the three-component system hexadecyltrimethylammonium bromide (CTAB)–hexanol–water. L_1, A region with water-rich solutions; L_2, a region with hexanol-rich solutions; D and E, lamellar and hexagonal liquid crystalline phases, respectively. Also indicated are the structures of normal (L_1 region) and reversed (L_2 region) micelles, as well as of the liquid crystalline phases (from Lindman and Forsén, 1976.)

tion of amphiphiles makes the lamellar phase of simple systems an appropriate model system for various aspects of cell membranes. For reviews of the field of surfactant and biological amphiphilic systems, a number of different articles will be found useful (Ekwall, 1975; Lindman and Wennerström, 1980; Mittal, 1977; Mukerjee and Mysels, 1971; Tanford, 1973). The multinuclear NMR approach to surfactant systems is illustrated by the spectra in Fig. 2.

As can be seen, there are a variety of geometries for the polycharged surfaces—spheres (convex and concave), cylinders (convex and concave), planes, and others. It may be noted that these are idealized geometries describing only approximately the actual situation. An important generalization from a number of experimental and theoretical studies is that many important features are qualitatively invariant among different geometries.

II. Counterion Binding: Methods and Theories

The high density of charged groups on the surface has important consequences for the behavior of these systems. The high total charges produce

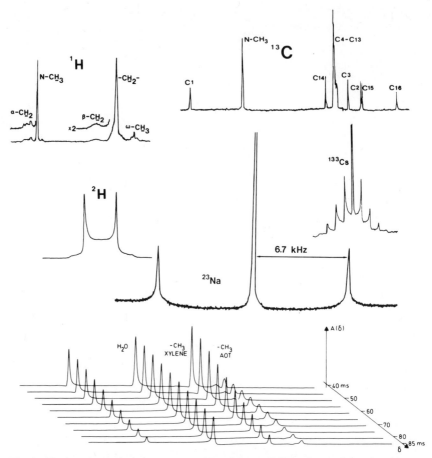

Fig. 2 Nuclear magnetic resonance provides rich possibilities for studying the structure and dynamics of surfactant systems. ^{1}H and ^{13}C spectra for micellar solutions of hexadecyltrimethylammonium bromide and ^{2}H (of heavy water), ^{23}Na, and ^{133}Cs quadrupole split spectra of mesomorphous phases. At the bottom is a ^{1}H FT PGSE experiment with a microemulsion. AOT denotes Aerosol OT i.e. sodium di-2-ethylhexylsulfosuccinate. (By courtesy of A. Khan and P. Stilbs).

very strong interparticle (polymer, micelle, etc.) repulsions keeping the particles far apart and preventing association. Typically, these repulsions are so great that a lattice-like arrangement of the particles (detectable, for example, in X-ray studies) is found. For flexible systems, the strong intergroup repulsions on the surface forcing the charged groups apart have important manifestations, such as the spherical shape of first-formed micelles. However, the most important aspect and the one we shall be mainly concerned with here is the influence of the high charge densities on small counterions. To reduce the energetically unfavorable high charge densities, the counterion distribution is perturbed so that the concentration is much higher in the vicinity of the polyion surface than in the bulk. One therefore speaks of counterion binding, the characteristics of which have been well illustrated in NMR studies and will be described here. It should be stressed at the outset that the interaction of small counterions with polyions deviates in basic respects from what is conventionally termed binding in chemistry. As will be demonstrated, the counterion–polyion association typically does not result in well-defined stoichiometric complexes involving short-range forces, but rather is a result of long-range electrostatic interactions. The interaction of the counterions with the polyion as a whole rather than with single groups has important consequences for the basic interpretation of experimental results and the choice of methodology. As a knowledge of these aspects is a prerequisite in using the various applications of NMR, a brief introduction is appropriate.

The possibilities of monitoring the interaction of small ions with polyelectrolytes in experimental studies are numerous, but the experimental methods can be divided into three groups depending on which aspect of the counterion binding is being investigated (Gunnarsson *et al.*, 1980). These groups are as follows.

(a) *Thermodynamic.* Examples are potentiometric studies on the counterion activity with various ion-specific electrodes or measurements of osmotic pressure. Thermodynamic methods monitor the counterion concentration far from the micelle. The degree to which it is decreased in comparison with the total counterion concentration characterizes the counterion binding.

(b) *Transport.* Examples are conductance and diffusion studies. In tracer or NMR self-diffusion measurements, for example, one monitors the fraction of counterions moving with the polyion as a kinetic entity. In theoretical calculations, one may consider (somewhat arbitrarily) counterions bound if they are attracted by the polyion by more than the thermal energy kT.

(c) *Spectroscopic.* In spectroscopic studies, such as investigations of NMR chemical shifts, relaxation rates, and quadrupole splittings, one monitors the fraction of ions that have their spectroscopic properties appreciably affected by the polyions. The various NMR parameters are dominated by short-range effects, and therefore only counterions in close contact (typically separated by at most one solvent molecule) with the polyion are considered bound. From the counterion distribution curves obtained in theoretical calculations, one may consider counterions bound if they are within 3 Å of the polyion. This figure is rather ambiguous, but the general features are affected only a little by moderate changes in this value.

Because of the continuous counterion distribution, any distinction between bound and free counterions is necessarily artificial. However, both experimental and theoretical studies indicate that the two-site model (as indicated previously for the different experimental approaches) provides a rather good characterization of the counterion–polyion interaction. It is therefore useful to introduce the degree of counterion binding β, defined as the ratio of bound counterions (in equivalents) to the total number of charged groups on the polyion. It should be realized that β depends on the type of experimental approach (which may be indicated by a subscript, e.g., β_{th}, β_{tr}, or β_{sp}) and that it does not give a complete characterization of the counterion distribution.

Theoretical models of the counterion–polyion interaction either describe the association in terms of a number of equilibria with individual polyion groups or consider the electrostatic interactions between small counterions and an idealized geometry of smeared-out charge representing the polyion. In the former case, applying the law of mass action leads to considerable difficulty. This treatment of chemical equilibria is appropriate for cases of chemical association due to short-range interactions. Attempts have been made to take long-range electrostatic interactions into account with various corrections such as site–site interaction constants and activity coefficients. However, these approaches have proved to be unsatisfactory and do not provide a good understanding of the system. In biochemistry, for example, it was rather common in earlier work to make an arbitrary division into two or more classes of binding sites when in fact typical polyelectrolyte effects were the explanation. The early treatments have now been superseded by models that explicitly consider long-range electrostatic interactions. The most important models are the ion condensation model, the Poisson–Boltzmann (PB) model, and Monte Carlo simulations.

In the ion condensation model (Oosawa, 1971; Manning, 1978), which has been derived for linear macromolecules, the polyion is described as a uniform line charge. According to this model, no counterions are bound if the charge density is below a certain critical value, whereas above this value counterions are bound to the extent that they reduce the effective charge density to the critical value. The charge density parameter is defined as

$$\xi = e^2/4\pi\varepsilon\varepsilon_0 kTb = 7.1/b \tag{1}$$

where $\varepsilon\varepsilon_0$ is the permittivity of the solvent and b the distance (in angstroms) between adjacent charges along the linear extension of the polyion; e, k, and T have their usual meanings. For water at 25°C, $e^2/4\pi\varepsilon\varepsilon_0 kT = 7.1$ Å. Counterion condensation occurs when $\xi > |Z|^{-1}$, where Z is the counterion valency. For univalent ions, condensation thus occurs when the (unneutralized) charges are closer to each other than 7.1 Å, and for bivalent ions the critical distance is 14.2 Å. Without added salt, the fraction of bound (condensed) counterions P_b is predicted to be

$$P_b = 1 - 1/\xi|Z| \tag{2}$$

The major limitations of the ion condensation model are that it considers only one geometry (linear) and that it is unrealistic in neglecting the radius of the polyelectrolyte. It seems more reasonable to treat the polyion as a uniformly charged three-dimensional body, e.g., a sphere, cylinder, or plane. If the mobile ions are treated as uncorrelated point charges, the distribution of counterions outside the polyion may be obtained as a solution of the PB equation (see e.g., Lifson and Katchalsky, 1954; Dolar and Peterlin, 1969; Engström and Wennerström, 1978; Fixman, 1979; Guéron and Weisbuch, 1980; Jönsson et al., 1980; Jönsson, 1981)

$$-\varepsilon\varepsilon_0 \left(\frac{\partial^2\psi}{\partial x^2} + \frac{\partial^2\psi}{\partial y^2} + \frac{\partial^2\psi}{\partial z^2} \right) = \rho = \sum eZ_iC_i = \sum eZ_iC_{i0} \exp\left(-eZ_i\psi/kT\right) \tag{3}$$

where ψ and ρ are the electric potential and the charge density, respectively, C_i the concentration of ion species i, and C_{i0} a normalization constant with the dimension of concentration. Wennerström and co-workers, for example [see previous references and the review by Lindman and Wennerström (1980)], have solved the PB equation for different geometries and find that, above a certain ionization of the polyion, the free counterion concentration changes very little with ionization whether the body is a sphere, a cylinder, or a plane. The effective surface charge density rapidly attains saturation values and, furthermore, the fraction of bound ions changes very little even on strong dilution of the system. These calculations thus suggest an approximate ion condensation behavior for all geometries.

Several attempts to overcome some limitations of the PB approach, such as the assumption of a medium with a fixed dielectric constant and the neglect of ion–ion correlations and of the ionic radius, have been made using modified PB equations (Outhwaite, 1970; Levine and Outhwaite, 1978; Bhuiyan et al., 1981; Jönsson, 1981) and performing Monte Carlo simulations; (Valleau and Torrie, 1982; van Megen and Snook, 1980). It is evident that Monte Carlo simulations are promising and will greatly increase our understanding of polyelectrolyte systems. With particular emphasis on surfactant systems it has been shown that the PB equation describes the ion distribution well for univalent counterions (for moderate salt concentrations) but not for bivalent ones (Jönsson *et al.*, 1980; Linse and Jönsson, 1982; Linse *et al.*, 1982).

III. Counterion Binding: Quadrupole Splittings

A large number of common counterions have sizable nuclear quadrupole moments, and quadrupole interactions are generally the dominant and most interesting aspect of counterion NMR in polyion systems. Although for simple solutions of small kinetic entities the NMR signals of counterions generally are single Lorentzians and thus relatively poor in information, more complex features are encountered in polyion systems. The following situations will mainly be considered.

(a) For anisotropic systems, mainly mesomorphous amphiphilic systems but also oriented polymer systems, quadrupole splittings are observed (Fig. 2). Generally first-order effects are encountered, but shifts due to second-order effects are also common for nuclei with large quadrupole moments. The quadrupole splittings give information on orientation effects, both at the microscopic and at the macroscopic levels, as well as on the degree and mode of ion binding (Lindblom *et al.*, 1976).

(b) For isotropic macromolecular solutions, relaxation is generally nonexponential and the decays of the longitudinal and transverse magnetizations differ. Relaxation depends on the degree and mode of ion binding and on the correlation times of the motions causing relaxation (Forsén and Lindman, 1981; Lindman and Forsén, 1976).

The NMR spectrum for an anisotropic system consists of $2I$ equidistant peaks, and we define the quadrupole splitting (Δ) as the distance (in frequency units) between two adjacent peaks. Rather detailed presentations of the theory and principles may be found elsewhere (Lindblom *et al.*, 1976), and here we shall present only some useful equations.

The counterions usually exchange rapidly between different environments in the system, and observed quadrupole splittings or relaxation rates are population-weighted averages over the different sites. For simplicity, we consider here only a two-site situation with free (f) and bound (b) counterions. Then the (first-order) quadrupole splitting may be written

$$\Delta(\theta) = [3/4I(2I - 1)]|(3 \cos^2 \theta - 1)(P_b S_b \chi_b + P_f S_f \chi_f)| \tag{4}$$

for a macroscopically oriented sample having the phase symmetry axis at an angle of θ to the magnetic field. $\chi = e^2qQ/h$ is the quadrupole coupling constant and S the order parameter. Generally, the second term may be neglected, because S_f is close to zero. For a polycrystalline or powder sample, the NMR spectrum is a superposition of spectra for different θ's. The frequency separation between adjacent peaks is given by (assuming $S_f = 0$)

$$\Delta_b = [3/4I(2I - 1)]P_b|S_b\chi_b| \tag{5}$$

Unfortunately, it is not easy to separate the three factors determining the quadrupole splitting, all of which provide information on the system.

IV. Counterion Binding: Quadrupole Relaxation

The quadrupole relaxation rate of ions (or of water molecules) differs appreciably in aqueous salt solutions and aqueous solutions containing polycharged species—the difference being mainly in the form of the spectral density (Forsén and Lindman, 1981; Lindman and Forsén, 1976). Two important problems in interpreting quadrupole relaxation data relate to the relaxation behavior of high-spin nuclei under nonextreme narrowing conditions and to the functional form of the spectral density for the quadrupole interaction. In regards to the former problem, it should be recalled that, for $I = \frac{3}{2}, \frac{5}{2}, \frac{7}{2},...$, both transverse and longitudinal magnetization decay nonexponentially, each being given by a sum of $I + \frac{1}{2}$ exponential decays. For $I = \frac{3}{2}$, the nonexponentiality can be easily observed and be very useful, whereas it is less important and often negligible for $I = \frac{5}{2}$ and $I = \frac{7}{2}$ (Bull, 1972; Bull et al., 1979; Halle and Wennerström, 1981a). In regard to the spectral density, we shall briefly consider the two-step model of relaxation (Halle and Wennerström, 1981b; Wennerström et al., 1974).

In considering the general features of relaxation, let us take $I = \frac{3}{2}$ as an example. The longitudinal relaxation is described by

$$M_L(t) - M_{L0} = [M_L(0) - M_{L0}](0.2e^{-a_1t} + 0.8e^{-a_2t}) \tag{6}$$

and the transverse magnetization by

$$M_T(t) = M_T(0)(0.6e^{-b_1t} + 0.4e^{-b_2t}) \tag{7}$$

The four relaxation rates a_1, a_2, b_1, and b_2 are given by

$$a_1 = (2\pi^2/5)\chi^2\tau_c/(1 + \omega^2\tau_c^2) \tag{8}$$

$$a_2 = (2\pi^2/5)\chi^2\tau_c/(1 + 4\omega^2\tau_c^2) \tag{9}$$

$$b_1 = (\pi^2/5)\chi^2[\tau_c + \tau_c/(1 + \omega^2\tau_c^2)] \tag{10}$$

$$b_2 = (\pi^2/5)\chi^2[\tau_c/(1 + \omega^2\tau_c^2) + \tau_c/(1 + 4\omega^2\tau_c^2)] \tag{11}$$

For the longitudinal relaxation it is difficult to determine the two relaxation rates, but the transverse relaxation often can be analyzed in terms of one rapidly relaxing component, accounting for 60% of the intensity, and one slowly relaxing one. It can be inferred that knowing any two of the relaxation rates permits the determination of both χ and τ_c when motion can be described simply in terms of a single correlation time (Forsén and Lindman, 1981). If one is relatively close to the extreme narrowing limit ($\omega\tau_c \lesssim 1.5$), linearized expressions of $1/T_1$ and $1/T_2$ may be employed (Bull, 1972; Halle and Wennerström, 1981a).

In practice, there are usually at least two sites with different relaxation properties to be taken into account. The equations given then apply for the macromolecular or bound site, and for the free counterions the extreme narrowing value $(2\pi^2/5)\chi^2\tau_c$ applies. In the limit that chemical exchange proceeds much more rapidly than a_1, a_2, b_1, and b_2, the observed four relaxation rates are simple population-weighted averages as usual. If the distribution of counterions between the sites is known, an analysis may be directly performed as previously indicated. In the limit of slow exchange, one narrow signal from the free ions and a broad (nonexponential) signal from the bound ions are observed. The latter is difficult to detect if the quadrupole interaction is strong but, if observable, yields direct information on χ and τ_c for bound ions. The behavior becomes more complex, however, for intermediate-exchange rates. Pertinent equations as well as procedures for establishing which exchange situation is at hand may be found elsewhere (Bull, 1972; Forsén and Lindman, 1981; Lindman and Forsén, 1976; Westlund and Wennerström, 1982).

The observation of quite small (compared to a rigid lattice) quadrupole splittings (and other static NMR effects) for amphiphilic liquid crystals and oriented polymers demonstrates the existence of rapid local motions of ions and water molecules at the charged interface. It is now well appreciated that, at the molecular level (structurally and dynamically), these anisotropic systems differ rather little from the corresponding isotropic ones. Such arguments led to the development of a two-step model for relaxation (Halle and Wennerström, 1981b; Wennerström et al., 1974). This model has been used rather extensively in NMR studies on, inter alia, surfactant micelles and microemulsions with ^{13}C, ^{1}H, ^{23}Na, etc., and to some extent also on macromolecular (polyelectrolyte and protein) systems.

A thorough treatment of NMR relaxation in colloidal and heterogeneous systems has been presented by Halle and Wennerström (1981b). Starting from the local anisotropy, two motional components are considered—a fast anisotropic motion superposed on a more extensive slow motion. It has been experimentally verified for several cases that these motions occur on widely different time scales and that the slow motion gives sizable contributions to relaxation. This two-step model of relaxation provides convenient expressions for the various relaxation rates. Relaxation is described in terms of three quantities:

(a)   One correlation time (or correlation function) for the rapid local motion τ_c^f. This time constant may represent, for counterions, motions in the primary hydration shell (Engström *et al.*, 1982) or, for water molecules, their (restricted) reorientation.

(b)   The residual interaction after partial averaging by the rapid motions is given by an order parameter S. If the local molecular structure at the charged interface is the same in the isotropic phase as in the corresponding anisotropic phase, quadrupole splittings in the latter give the same order parameter. This observation can be helpful in data analysis.

(c)   One correlation time (or correlation function) for the slow overall motion τ_c^s. Depending on system geometry, etc., the slow motion may be determined by different motions and may not be describable in terms of a single correlation time. For the simplest case of a spherically symmetric polyion, the residual interaction may be modulated by polyion reorientation τ_c^r, by lateral diffusion on the polyion surface τ_c^d, and by chemical exchange with the bulk solution τ_c^{ex}. Then $(\tau_c^s)^{-1} = (\tau_c^r)^{-1} + (\tau_c^d)^{-1} + (\tau_c^{ex})^{-1}$. For nonspherical cases, it may be necessary to describe each of these processes by more than one correlation time. Typically, however, for polyion and amphiphile systems, very different time scales are involved, so that the interpretation is relatively straightforward.

To analyze the data according to the two-step model, and thus to determine τ_c^f, S, and τ_c^s, one needs three independent relaxation rates. One possibility is to determine T_1 at three different magnetic fields, and another is to combine T_1 and T_2 data. One may also make use of the nonexponential relaxation of, for example, spin-$\frac{3}{2}$ nuclei.

V. Quadrupole Relaxation Studies on Counterion Binding to Polyelectrolytes

Quadrupole relaxation studies on counterion binding to polyelectrolytes have mainly concerned Na^+, but there are also a number of studies on K^+, Cl^-, Mg^{2+}, Ca^{2+}, etc. As an example, we shall describe investigations using ^{23}Na NMR. These results are in many respects rather straightforward to interpret because rapid exchange conditions normally apply. The relaxation behavior of spin-$\frac{5}{2}$ or $\frac{7}{2}$ nuclei under intermediate exchange conditions is quite complex, and no theory has been available. Therefore, it has so far been impossible to interpret adequately ^{25}Mg and ^{43}Ca relaxation results. However, as a result of recent theoretical work by Westlund and Wennerström (1982), this situation is now changing.

In studies on Na$^+$ binding to PMA of different ionization degrees α, typical nonextreme narrowing conditions were observed; i.e., T_1 and T_2 were unequal, transverse relaxation was biexponential, and spectra were non-Lorentzian (Gustavsson *et al.*, 1978a). These features are illustrated in Figs. 3–5. An analysis of these data gave the correlation time for the slow motion and on the basis of estimated degrees of counterion binding also the effective quadrupole coupling constant. As can be inferred from Fig. 6, τ_c^s varies considerably with the degree of ionization α. The values of the correlation time, as well as the variation with the charge density, are incompatible with the slowest motion associated with the reorientation of the (approximately cylindrical) polyion. Therefore, the slow-motion correlation time appears to be associated with either the lifetime of a counterion at the polyion or internal motions within the polyion. The counterion binding to PMA is relatively complex, with an effective quadrupole coupling constant that varies from 300 to 100 kHz as α is increased. The relatively low values support the assumption that a major part of the quadrupole interaction is modulated by rapid local motions. The mechanism behind the α dependence of χ (or, better, $S\chi$) is not clear. ^{23}Na chemical shifts also demonstrate a distinct change in counterion binding, and from a sign change of the chemical shift (relative to the free aqueous

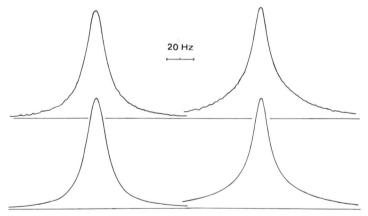

20 Hz

Fig. 3 Experimental (upper) and theoretical (lower) ^{23}Na spectra for 0.47 *m* aqueous PMA at $\alpha = 0.94$ and $\alpha = 0.06$. The theoretical curves are for $\alpha = 0.94$, a Lorentzian line with $1/T_2 = 55.0$ s^{-1}, and for $\alpha = 0.06$, the weighted sum of two Lorentzian lines with the relaxation rates 180 and 42 s^{-1} (Gustavsson *et al.*, 1978a).

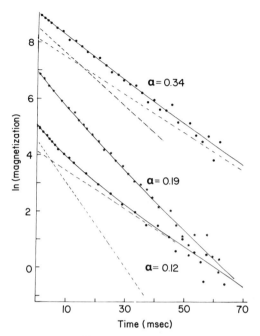

Fig. 4 The natural logarithm of ^{23}Na magnetization for aqueous PMA after a Carr–Purcell–Meiboom–Gill pulse sequence for three different α values. The dashed lines indicate the contribution to total magnetization from the slow and fast relaxing components (Gustavsson *et al.*, 1978a).

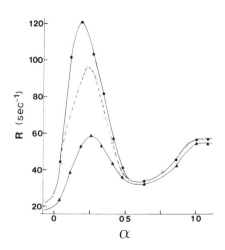

Fig. 5 ^{23}Na relaxation rates in reciprocal seconds for a 0.47 m aqueous PMA solution as a function of α. Here ● is R_2, ○ $\pi*\Delta\nu_{1/2}$, and ▲ R_1. (Gustavsson *et al.*, 1978a).

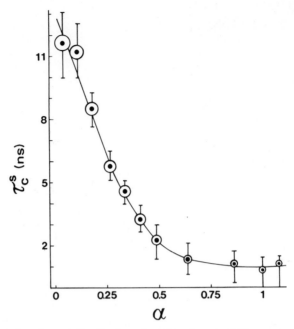

Fig. 6 Correlation time τ_c^s for the bound sodium ions in 0.47 m as a function of α (Gustavsson *et al.*, 1978a).

ion) it has been suggested that specific inter- or intrachain complexes $Na_{aq}^+-COO^--COOH$ are of importance. Such a chelation should retard segmental motion at low α values, but it should also influence the rate of Na^+ exchange.

Studies on counterion binding in a number of different polyelectrolyte systems have been presented by Leyte and co-workers. Kielman and Leyte (1973) found that T_1^{-1} for $^{23}Na^+$ at first increased with the degree of polymerization in aqueous solutions of polyphosphate and thereafter (above 60) became constant. For $^{23}Na^+$ in polyacrylate solutions, T_1^{-1} increases regularly with α (van der Klink *et al.*, 1974), and paramagnetic Mn^{2+} ions give a strongly enhanced relaxation (Westra and Leyte, 1979). Van der Klink *et al.* (1974) also made a thorough attempt to rationalize theoretically counterion quadrupole relaxation data for polyelectrolyte solutions. In this model, relaxation is due to translational motion of the counterion on an equipotential surface around the polyion treated as a rigid rod with a uniform surface charge density (Leyte 1979; van der Klink *et al.*, 1974). The results of Leyte and co-workers demonstrate that the PB model gives a much better description of counterion binding than the ion condensation model.

Recent ^{23}Na relaxation studies at very low polyelectrolyte concentrations (Levij et al., 1981; Levij et al., 1982) have demonstrated a dramatic increase in $\tau_c{}^s$ upon dilution. It has been suggested that this effect is correlated with conformational changes. An alternative explanation has been suggested, however (Halle and Wennerström, 1982). According to this view, the slowest correlation time in the system (i.e., the one that governs the rapidly relaxing component of the transverse magnetization) is, at low polyelectrolyte concentrations, the mean time required for a Na$^+$ ion to diffuse out of its polyion cell. Developing a multistep cell model of relaxation and calculating the mean diffusion time from the Smoluchowski equation, Halle and Wennerström (1982) obtained a quantitative rationalization of the low-concentration relaxation data.

There is still no general agreement about the basic aspects of the interpretation of counterion quadrupole relaxation in polyelectrolyte systems. However, the two-step relaxation model (Halle and Wennerström, 1981b) seems to be a reasonable first approximation that is in agreement with many observations. It has been pointed out (Gustavsson et al., 1978a; Meurer et al., 1978) and also theoretically demonstrated (Wennerström, 1980) that counterion diffusion around the polyion rods is not sufficient to average out the entire quadrupole interaction. This has been experimentally demonstrated through the observation of quadrupole splittings for oriented polyelectrolytes (Edzes et al., 1972) and through the deduction of correlation times considerably longer than that for diffusion around the polyion rod ($\sim 10^{-10}$ s) (Halle and Wennerström, 1982; Levij et al., 1981; Levij et al., 1982). In quantifying the field gradient fluctuations, reference may be made to a recent Monte Carlo study (Engström et al., 1982) which also examines a previously used polarization factor.

In regard to biological polyelectrolyte systems, much effort has been devoted to both polysaccharides (Grasdalen and Smidsrod, 1981; Gustavsson et al., 1978b; Gustavsson et al., 1981; Herwats et al., 1977) and nucleic acids (Anderson et al., 1978; Bleam et al., 1980; Reimarsson et al., 1979; Reuben et al., 1975; Rose et al., 1980). In a study on Na$^+$ binding to certain mucopolysaccharides a rather simple counterion binding pattern was encountered (Gustavsson et al., 1978b). The correlation time was obtained from T_1 and T_2 measurements using linearized expressions for the relaxation rates. A weak decrease in $\tau_c{}^s$ with increasing α was found. Based on the observed transverse relaxation rates and the deduced $\tau_c{}^s$ values and assuming constant values for the quadrupole coupling constants, the fraction of bound counterions was obtained. As shown in Fig. 7, P_b is low below $\alpha \simeq 0.65$ and then increases linearly with α to maximal ionization for both dermatan sulfate and chondroitin sulfate. This behavior is in qualitative agreement with the ion condensation model and the PB treatment.

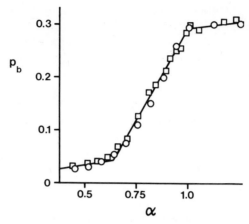

Fig. 7 Fraction of sodium ions bound P_B to dermatan sulfate (○) and chondroitin sulfate (□), obtained from ^{23}Na relaxation rates. Quadrupole coupling constants are 200 and 120 kHz for dermatan sulfate and chondroitin sulfate, respectively (Gustavsson *et al.*, 1978c).

Recent studies on polysaccharide systems involve more complex effects such as the binding of Na^+ ions to proteoglycans (Gustavsson *et al.*, 1981), the conformational behavior and ionization of hyaluronate (Piculell, 1982), the binding of iodide to κ-carrageenan (Grasdalen and Smidsröd, 1981), and the binding of Ca^{2+} and Mg^{2+} ions (Gustavsson, 1982). Na^+ binding to a chondroitin sulfate–peptide complex (CSP, a model of proteoglycans) is similar to that of simple mucopolysaccharides, except that there is a dramatic change in both τ_c^s and P_b at about $\alpha = 1$ (Fig. 8) (Gustavsson *et al.*, 1981). This appears to be associated with an aggregation of the polysaccharide chains. It was also demonstrated that the addition of Ca^{2+} ions enhanced Na^+ binding.

For κ-carrageenan, another polyanion system, Grasdalen and Smidsröd (1981) observed an interesting coion effect. Thus a specific effect of I^- in stiffening the polymer chains and thereby influencing intrachain conformations was investigated with ^{127}I NMR. A strong relaxation enhancement points to specific I^- binding. The effect is strongly temperature-dependent, does not occur with other more highly charged carrageenans, and is absent for Cl^-. The results suggest a specific binding site for I^- that gives rise to stabilization of an ordered conformation below ~46°C.

Using ^{23}Na, ^{25}Mg, and ^{17}O NMR (Fig. 9), Piculell (1982) demonstrated various aspects of the hydration and ion binding of hyaluronate. His results support the notion of a relatively rigid conformation at intermediate pH values but show above all, perhaps for the first time, that there are two distinct ionization steps. In qualitative agreement with the ion con-

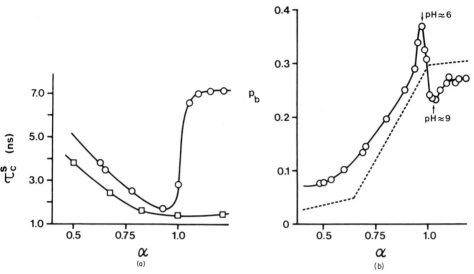

Fig. 8 (a) The variation in correlation time τ_c^s (nanoseconds) for bound Na^+ with the degree of neutralization α in chondroitin sulfate–peptide (CSP) (○) and dermatan sulfate (DS) (□) solutions. (b) The fraction P_b of Na^+ bound to 0.18 m aqueous CSP solution as a function of α. The dashed line gives the corresponding variation in P_b for pure CS and DS.

densation model, the first ionization produces binding mainly of bivalent counterions, whereas the second ionization gives a linear charge density high enough to produce binding of univalent counterions also.

The temperature dependence of the ^{25}Mg transverse relaxation in systems of Mg^{2+} and several different polyanions has been investigated by Gustavsson (1982). As shown in Fig. 10, a complex and widely varying pattern is found but, using the recent theory of Westlund and Wennerström (1982), these data can be well rationalized to provide, inter alia, the average lifetimes of Mg^{2+} ions at the polyions (Gustavsson, 1982).

VI. Quadrupole Relaxation and Chemical Shift Studies on Counterion Binding in Amphiphilic Systems

Investigations of various amphiphilic systems were probably the first examples of studies on counterion binding in macromolecular systems with the quadrupole relaxation method. In one study, it was demonstrated that the $^{81}Br^-$ quadrupole relaxation changed drastically on micellization

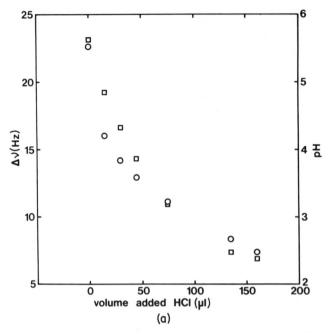

Fig. 9 (a) $^{25}Mg^{2+}$-hyaluronate. $^{25}Mg^{2+}$ line widths ($\square$, left scale) at 15.61 MHz and pH ($\bigcirc$, right scale) in a hyaluronate solution plotted against the volume of added 0.2 M HCl. The temperature was 22°C, and the initial concentrations in the 2.9-g sample were hyaluronate, 11 mmol (disaccharide)/kg H_2O; Na^+, 12 mmol/kg H_2O; and $^{25}MgCl_2$, 5 mmol/kg H_2O. The line width of a $^{25}MgCl_2$ reference solution, measured under the same conditions, was 5.2 Hz. (b) $^{23}Na^+$-hyaluronate. $^{23}Na^+$ excess longitudinal relaxation rates ($\square$, left scale) at 67.45 MHz and pH ($\bigcirc$, right scale) in a hyaluronate solution plotted against the volume of added 2 M HCl. The temperature was 22°C, and the initial concentrations in the 3.90-g sample were hyaluronate, 53 mmol (disaccharide)/kg H_2O; and Na^+, 138 mmol/kg H_2O. The excess longitudinal relaxation rate is defined as $R_{1,ex} = R_1 - R_{1,ref}$, where $R_{1,ref}$ is the measured longitudinal relaxation rate in a 0.11 M NaCl solution under the same conditions. (c) $^{23}Na^+$-chondroitin. $^{23}Na^+$ excess longitudinal relaxation rates ($\square$, left scale) at 67.45 MHz and pH ($\bigcirc$, right scale) in a chondroitin solution plotted against the volume of added 2 M HCl. The temperature was 22°C, and the initial concentrations in the 4.37 g sample were chondroitin, 60 mmol (disaccharide)/kg H_2O; and Na^+, 138 mmol/kg H_2O. (By courtesy of Lennart Piculell)

(Eriksson *et al.*, 1966), and another emphasized the insensitivity of Na^+ counterion binding to changes in amphiphile phase structure (and thus to major changes in aggregate geometry and size) (Lindman and Ekwall, 1968). Later work has confirmed and extended these observations and, using theoretical results from the work of Wennerström, has provided a rather detailed picture of counterion binding in surfactant and biological amphiphile systems. Above all, the results tend to emphasize the interac-

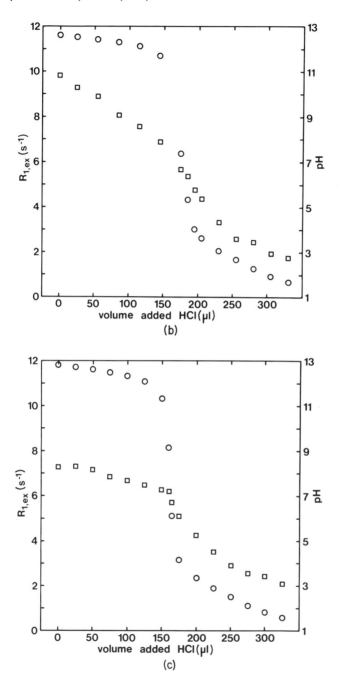

(b)

(c)

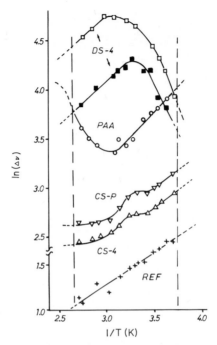

Fig. 10 The logarithm of the $^{25}Mg^{2+}$-NMR line width versus the inverse temperature for the following aqueous solutions: Dermatan 4-sulfate (DS-4), [DS-4] = 70 mM, [Mg^{2+}] = 181 mM, pH = 4.05 (□); [DS-4] = 68 mM, [Mg^{2+}] = 125 mM, pH = 3.17 (■). Polyacrylate (PAA), [PAA] = 93 mM, [Mg^{2+}] = 62 mM, pH = 4.19 (○). Chondroitin 4-sulfate (CS-4), [CS-4] = 70 mM, [Mg^{2+}] = 125 mM, pH = 5.03 (△). A protein–chondroitin sulfate complex (CS-P), [CS-P] = 75 mM, [Mg^{2+}] = 122 mM, pH = 5.17 (▽). A 100 mM MgCl$_2$ solution (REF., +). (By courtesy of Hans Gustavsson.)

tion of counterions with the amphiphile aggregate as a whole rather than the specific interaction with individual charged groups. An important complement to NMR relaxation and chemical shift studies is the investigation of counterion self-diffusion. This work has mainly been performed with radioactive labeling (Lindman and Brun, 1973; Lindman et al., 1982) and very little with NMR. However, recently the Fourier transform pulsed-gradient spin-echo NMR technique has been applied (Stilbs and Lindman, 1981) and will no doubt be very useful in future work.

As previously mentioned, it is common to discuss counterion binding to micelles (and also to other surfactant aggregates) in terms of degree of counterion binding β. It is directly observed in self-diffusion studies that β_{tr} is quite insensitive to changes in micelle concentration over a wide range, β lying in the range 0.5–0.8 (Lindman and Wennerström, 1980; Lindman et al., 1982). This is also directly borne out in studies on coun-

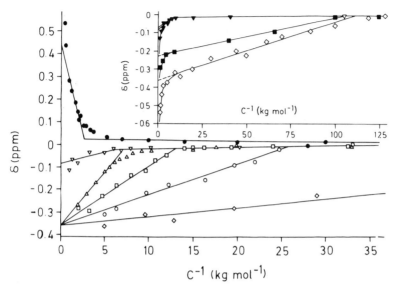

Fig. 11 ^{23}Na-NMR chemical shifts as a function of the inverse amphiphile concentration (molality), for aqueous solutions of sodium octanoate (●), octyl sulfonate (▼), octyl sulfate (△), nonyl sulfate (□), decyl sulfate (○), dodecyl sulfate (◇), and (in the inset) p-octylbenzene sulfonate (■). Straight lines according to predictions from the phase separation model of micelle formation (Gustavsson and Lindman, 1978).

terion chemical shifts and quadrupole relaxation rates for a large number of systems; such studies have been performed using, inter alia, ^{23}Na, ^{35}Cl, ^{85}Rb, ^{133}Cs, and ^{81}Br NMR (Gustavsson and Lindman, 1975; Gustavsson and Lindman, 1978; Robb and Smith, 1974; Wong *et al.*, 1977). The results obtained, as shown in Figs. 11 and 12, also demonstrate that the intrinsic chemical shifts and relaxation rates of bound counterions (δ_b and R_b) are nearly independent of micelle concentration and of the alkyl chain length of the surfactant. On the other hand, these quantities are dependent on the surfactant polar head group (Gustavsson and Lindman, 1978).

Efforts have been made to deduce information on the hydration state of bound counterions from various parameters. One possibility is to use deduced values of δ_b and R_b as a starting point, because transfer from an aqueous to a nonaqueous medium should produce quite sizable changes in both the paramagnetic shielding term and the electric field gradients. One infers directly from such comparisons that alkali counterions remain hydrated when bound to micelles (Lindman *et al.*, 1977). As an example, the change in ^{23}Na shielding on binding to a micelle (less than 1 ppm; see Fig. 11) is an order of magnitude smaller than the typical change on transfer from water to another solvent (on the order of 10 ppm: Lindman

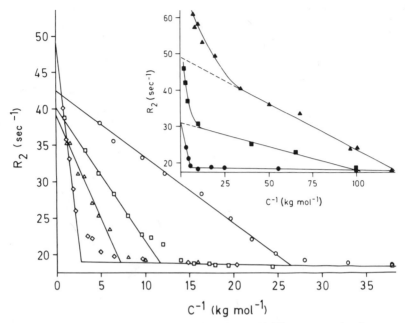

Fig. 12 ^{23}Na relaxation rates R_2 versus the inverse amphiphile concentration for aqueous solutions of the sodium salts of octanoate ($\Diamond$), octyl sulfate ($\triangle$), nonyl sulfate ($\square$), decyl sulfate ($\bigcirc$), dodecyl sulfate ($\blacktriangle$), octylbenzene sulfonate ($\blacksquare$), and octyl sulfonate ($\bullet$). The data for the last three substances are shown in the inset where some data for the high concentration region are also included. Straight lines according to predictions from the phase separation model (Gustavsson and Lindman, 1978).

and Forsén, 1978). A more direct demonstration is given by the water solvent isotope effect on the chemical shift as illustrated in Fig. 13 for ^{133}Cs$^+$ in cesium octanoate solutions. In fact, the difference in chemical shift in H$_2$O and D$_2$O solutions is, within experimental error, the same for concentrated micellar solutions as for infinite dilution (Gustavsson and Lindman, 1978).

For other phases of complex surfactant systems, the use of NMR is even more significant, because many other techniques are inapplicable for different reasons. It is a general observation that both chemical shift and quadrupole relaxation are rather insensitive to changes in phase and aggregate structure, for example, sphere → rod micelle → hexagonal liquid crystal → lamellar liquid crystal → reversed micelle. However, there is still a significant increase in counterion binding in this series. It appears that anions are somewhat more sensitive to these changes than cations, presumably as a result of more probable direct contacts with the polar heads. This approximate insensitivity of counterion binding was first rather surprising but has recently been rationalized from electrostatic

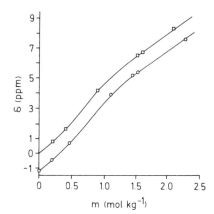

Fig. 13 ^{133}Cs chemical shifts of H_2O ($\square$) and D_2O ($\bigcirc$) solutions of cesium octanoate as a function of amphiphile molality. (Gustavsson and Lindman, 1978.)

calculations using the PB approach. These studies indicate that the general features of counterion binding can be understood by treating the water as a dielectric continuum, the amphiphilic aggregates as uniformly charged bodies, and the counterions as point charges (Gunnarsson *et al.*, 1980; Jönsson, 1981).

Much interest is presently focused on the counterion binding in various types of microemulsion systems. A number of NMR studies have been performed on L_2 phases containing micelles of the reversed type (Lindman and Ekwall, 1968; Rosenholm and Lindman, 1976; Sjöblom and Henriksson, 1982). It is then observed that, for the water-rich parts of the phases, counterion quadrupole relaxation is relatively slow and not much different from that for concentrated micellar solutions of the same system. However, as the water content is decreased, one observes a dramatic increase in relaxation. Generally, the increase starts in a rather narrow concentration range. It appears that the water/counterion molar ratio corresponding to this increase is correlated with counterion hydration. Also, chemical shift studies demonstrate a decreased water–counterion contact starting in a narrow concentration range (Gustavsson and Lindman, 1978). These various changes in counterion NMR demonstrate a transition from reversed micelles to less associated solutions where the counterions appear in ion pairs.

VII. Quadrupole Splitting Studies on Counterion Binding in Amphiphilic Systems

The feasibility of observing counterion quadrupole splittings from amphiphilic mesophases was first demonstrated by Lindblom (Lindblom *et*

al., 1971; Lindblom, 1971; Lindblom, 1972) about 10 yr ago in studies on
^{35}Cl, ^{37}Cl, and ^{23}Na; for Cl, the quadrupole effects were so large that only
second-order effects were discernible, which complicated the investigations. Lindblom *et al.* extended the studies during the following years to
cover a wide range of phenomena and systems, and Wennerström's theoretical work provided a basis for interpretation (Lindblom and Lindman,
1973; Lindblom *et al.*, 1976; Lindblom *et al.*, 1978; Wennerström *et al.*,
1974; Wennerström *et al.*, 1979). Furthermore, several other groups have
started similar investigations (Abdolall *et al.*, 1977; Charvolin *et al.*, 1977;
Loewenstein *et al.*, 1978; Reeves, 1975). Counterion quadrupole splitting
studies may be performed to investigate different aspects of amphiphilic
systems, such as phase structure and homogeneity and macroscopic
alignment, but in general these aspects are better studied using ^{2}H NMR
(Section IX).

The most important aspect of counterion quadrupole splitting studies is
that they provide unique insight into ion binding. As previously discussed
[Eq. (5)], quadrupole splittings are usually determined by three factors,
P_b, S_b, and χ_b, which are not easily separated. Fortunately, it appears
that, for a wide range of systems, S_b and χ_b are independent of composition, so that the quadrupole splittings can be used conveniently to monitor
changes in the degree of counterion binding (Lindblom *et al.*, 1976). However, there is at least one situation where S_b is very sensitive to composition and temperature and even changes sign, and this is for alkali ion
binding in systems with carboxylate groups. This complex situation, giving rise to difficulties in interpretation but also to an insight into the
location of counterions at the amphiphile–water interface, cannot be adequately described here, but we refer the reader to the original work
(Lindblom *et al.*, 1975; Lindblom *et al.*, 1978). Instead, we consider the
simpler and more typical situation where Δ is roughly proportional to P_b.

Early studies suggested a marked invariance of counterion binding with
changes in temperature and concentration. Typical illustrations of this are
given in Figs. 14 and 15 for two rather different systems. Such findings
were rather surprising based on the previous view of counterion binding,
and another surprising observation is presented in Fig. 16. It thus appears
that, on addition of an electrolyte containing counterions to a lamellar
phase, there is no change in the amount (or number) of bound counterions
but only in the amount of free ions. All these observations correspond to
ion condensation behavior as deduced theoretically from the PB equation,
as previously mentioned. Quadrupole splitting studies are apparently important in testing theoretical models of counterion binding.

Another aspect of counterion binding in amphiphilic mesophases that
may be conveniently studied in this way concerns ion competition. Com-

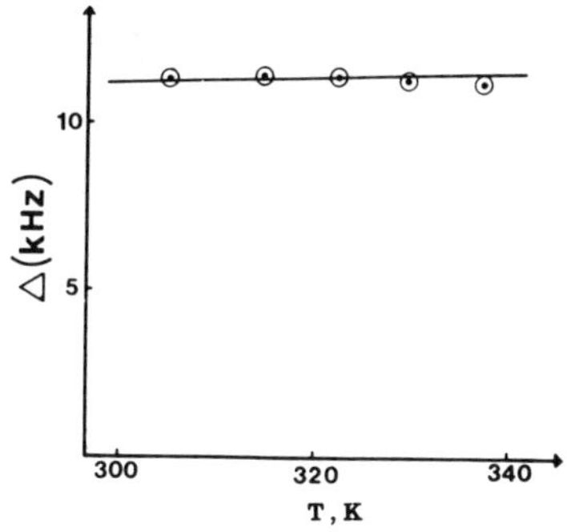

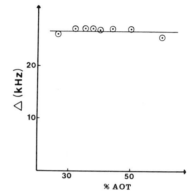

Fig. 14 Second-order ^{35}Cl splittings for the lamellar phase of the system water–octylammonium chloride–decanol as a function of the temperature. Circles represent experimental points. The solid line indicates the calculated temperature dependence based on the electrostatic theory (Wennerström et al., 1980).

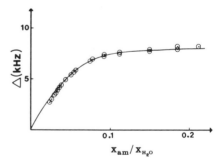

Fig. 15 The ^{23}Na quadrupole splitting as a function of the amphiphile concentration for the lamellar phase of the binary Aerosol OT–water system (Wennerström et al., 1980).

Fig. 16 The ^{23}Na splitting plotted against the ratio of total amphiphile to water for the lamellar phase of the system water–sodium dodecyl sulfate–hexanol–sodium chloride. The sodium chloride/water ratio was kept constant. ⊙, Experimental points. The solid line was calculated using the assumption that the amount of bound ions was constant (Wennerström et al., 1979; Wennerström et al., 1980).

petition both among and between alkali ions and calcium ions has been studied. Söderman et al. (1980) investigated counterion competition in a lamellar phase composed of water, decanol, and varying mixtures of two alkali octanoates with ^{7}Li, ^{23}Na, and ^{133}Cs NMR. The measurements showed that, when varying the ratio between the ions, the splitting for one ion increases, whereas it decreases for the other; this is as expected for

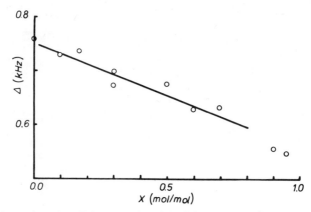

Fig. 17 ^{23}Na quadrupole splittings as a function of the ratio of Ca^{2+} to Na$^+$ surfactant (in equivalents) for the hexagonal phase of the system water–calcium octyl sulfate–sodium octyl sulfate. Sample composition (w/w): H$_2$O 41% and surfactants 59%. The solid line shows the theoretical curve obtained from PB calculations (Lindman *et al.*, 1980b).

competition behavior. It was observed that smaller alkali ions had a higher affinity for binding. The results are in qualitative agreement with PB calculations taking ion specificity into account simply by introducing a distance of closest approach between counterions and the lamellar surface. The results support the view that alkali ions bind hydrated. A study on Na$^+$–Ca^{2+} competition in the hexagonal phase of the system octyl sulfate–water gave similar results (Lindman *et al.*, 1980b). Thus good agreement with the PB calculations was found (Fig. 17), and it could be argued that counterions bind hydrated and do not bind to specific sites.

VIII. Hydration in Colloidal Systems: General Aspects and NMR Methods

Knowledge of the hydration of macroscopic and microscopic interfaces is a prerequisite for understanding the fundamental aspects of many chemical and biological systems, yet there is considerable controversy in the field of hydration. This situation arises both from inadequate application of experimental methods and from the use of inappropriate models in interpreting the results. When water is the solvent or dispersion medium, it is not straightforward (and perhaps not even meaningful) to make a distinction between free and bound water molecules, which is the general basis of assigning hydration numbers. In any case, the division into free and bound water molecules may differ considerably depending on the physicochemical property monitored. Consequently, the hydration number also varies.

Nuclear magnetic resonance investigations have contributed a great deal to the controversy in this field. It would carry us too far afield to discuss the various reasons here, but they include neglect of important relaxation effects in ${}^1H$ NMR, neglect of the effect of chemical exchange with prototropic groups in ${}^2H$ NMR, and the assumption in several relaxation studies (${}^1H$, ${}^2H$, and ${}^{17}O$ NMR) that water molecules bind rigidly at interfaces (Halle, 1981; Halle, 1982). The difficulty with ${}^1H$ and ${}^2H$ NMR in studying hydration and water dynamics led Halle to investigate in detail the possibilities of ${}^{17}O$-NMR relaxation. Halle (1981) demonstrated that in interpreting multifield relaxation data using the two-step model of relaxation it was indeed possible to map in detail the hydration of a large number of systems. The work of Halle and co-workers includes studies on micelles, microemulsions, polyelectrolytes, proteins, and colloidal suspensions, and a brief review is given in Section XI.

For anisotropic systems, the use of static NMR effects in the spectra of water nuclei offers a convenient way of monitoring hydration that is relatively free of interpretation difficulties. Proton exchange is a problem, but much easier to handle than in relaxation studies. Some applications of ${}^2H$ and ${}^{17}O$ quadrupole splittings in studying hydration are presented in Section X, and in Section IX we outline the use of ${}^2H$ NMR in investigating phase equilibria.

Self-diffusion studies constitute a further rather general approach to hydration, which is applicable to (relatively dilute) isotropic systems (Lindman et al., 1980a). A number of other approaches applicable to special systems and to more specific questions include studies on counterion hydration using the hydrogen–deuterium water isotope effect in ion chemical shifts (Section VI), studies on chain hydration using the hydrogen–deuterium water isotope effect on relaxation (dipolar relaxation) (Ulmius and Lindman, 1981), and studies on the effect of paramagnetic counterions (Cabane, 1981).

IX. Deuteron NMR for Phase Diagram Determinations in Surfactant Systems

Fundamental information concerning a surfactant system involves the identity of the phases formed and their stability ranges as a function of composition and temperature. The determination of a complete phase diagram is normally a laborious and difficult task because of the large number of phases that occur. However, complete phase diagrams for a large number of systems are available (Ekwall, 1975). Techniques employed in studying phase diagrams normally require macroscopic separation of the different phases in equilibrium, but the high viscosities and low density differences often cause great problems. In many instances, NMR can be a simple, useful tool in phase diagram determinations that does not require macroscopic separation of the phases.

Because the NMR spectra of quadrupolar nuclei have a completely different appearance for isotropic and anisotropic phases, their application is straightforward. As an example, a disputed phase (C phase) can easily be demonstrated not to be a single phase but a dispersion of liquid crystal and an isotropic solution, because it gives both a singlet and a doublet in ^{2}H NMR of D_2O (Persson et al., 1975). In an elegant study, the important temperature–composition phase diagram for dipalmitoyl lecithin–water was worked out in the same way (Ulmius et al., 1977). Here we shall illustrate the principles of this method based on a recent study where the entire phase diagram for a three-component system was obtained using ^{2}H NMR of D_2O (Khan et al., 1982b). The system calcium octyl sulfate–decanol–water is also of interest, because it was the first calcium surfactant system for which a phase diagram was obtained.

The principles of the technique are best discussed in reference to the experimental spectra of different phase regions shown in Fig. 18. Two simple features of NMR quadrupole splittings constitute the basis of the method. First, an isotropic phase gives only a singlet, whereas an anisotropic one gives a doublet with ^{2}H NMR. Second, the quadrupole splitting of a lamellar phase is expected to be much larger than that of a hexagonal one. Ideally, for identical local molecular effects, symmetry considerations predict twice as large a quadrupole splitting for a lamellar phase. In fact, this simple relation holds approximately for many cases but can be violated if the two phases have very different water contents.

Because water exchange between different phases occurs slowly, in a multiphase case signals from different phases appear superimposed on each other. It is therefore in general possible to determine directly from a spectrum how many (one, two, or three) phases are present and (as mentioned previously) which ones they are. The identification of three-phase triangles is particularly important in establishing the phase diagram. Problems arise in two cases, namely, when there is more than one isotropic phase and when the microcrystallites of a liquid crystal are so small that no splitting appears. (The lifetime in a microcrystallite is too short, and diffusion between microcrystallites averages out the quadrupole interaction.) However, as described by Khan et al. (1982b), these difficulties may be overcome.

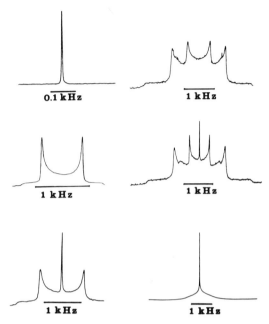

Fig. 18 Typical ^{2}H-NMR spectra of samples from different parts of the phase diagram for the calcium surfactant system (Khan *et al.,* 1982b).

Five different phases were established for this calcium surfactant system (Fig. 19), namely, isotropic aqueous and decanolic solution phases, normal hexagonal and lamellar liquid crystalline phases, and a solid hydrated surfactant phase. The phase diagram is quite similar to that for the corresponding system with sodium as the counterion, with one important exception; namely, the extension of the lamellar phase is very much smaller for the calcium system. It has been demonstrated that the ability of a lamellar phase to swell and incorporate water results from a repulsion between the amphiphilic layers. A much lower electrostatic repulsion and a much higher degree of counterion binding with calcium are in agreement with recent theoretical PB and, especially, Monte Carlo studies (Wennerström *et al.,* 1982).

Although ^{2}H NMR is the most generally applicable NMR method for phase diagram determination, other nuclei such as ^{19}F, ^{14}N, and ^{31}P may also be very useful (Arnold *et al.,* 1981; Khan *et al.,* 1980; Khan *et al.,* 1982a). ^{19}F- and ^{31}P-NMR studies are based on chemical shift anisotropy and its change in sign on going from lamellar to hexagonal liquid crystal.

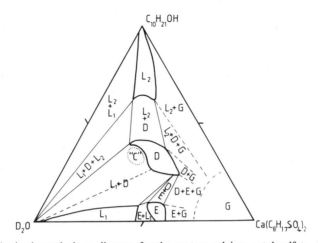

Fig. 19 The isothermal phase diagram for the system calcium octyl sulfate–decan-1-ol–heavy water at 27°C. L_1 and L_2, Isotropic aqueous and decanolic solutions, respectively; D, liquid crystalline phase with lamellar structure; "C", part of the two-phase zone $L_1 + D$ where the samples have the microscopic appearance of samples from the C region in the classic model system sodium octanoate–decan-1-ol–water; E, liquid crystalline phase with hexagonal structure; $L_1 + L_2$, $L_1 + L_2 + D$, etc., two and three-phase zones (Khan *et al.*, 1982b)

X. ²H- and ¹⁷O-NMR Quadrupole Splittings and Hydration in Amphiphilic Mesophases

Quadrupole effects in the NMR of water nuclei can be used to study hydration phenomena in exactly the same way that those of ions can be used to study counterion binding. It is believed that ²H- and ¹⁷O-NMR are rather unique in providing a detailed molecular picture of the hydration in these types of systems. For water molecules, the principles and equations are analogous to those for counterions (Wennerström *et al.*, 1975). The water exchange between different sites is (in all cases investigated so far) rapid, and thus a powder splitting is given by Eq. (5) if we make the same assumptions about a division into only two sites and about a negligible order parameter for free water molecules. With respect to the counterion studies, there are two important differences:

(a) A favorable one is that, because the field gradients are mainly of intramolecular origin and are little influenced by changes in the systems, the quadrupole coupling constants are known to a good approximation (0.222 MHz for ²H and 6.7 MHz for ¹⁷O) (Halle and Wennerström, 1981b). This facilitates considerably the interpretation of experimental data.

(b)   An unfavorable one is the greater difficulty in making a distinction between free and bound water molecules. A macroscopic or microscopic surface induces a perturbation of the water solvent, which decays gradually with the distance from the interface. However, NMR is most appropriate for monitoring this perturbation. In quadrupole splitting studies, water molecules that are appreciably oriented with respect to the amphiphile–water interface are considered bound.

The swelling behavior of lamellar phases in surfactant and biological amphiphile (mainly phospholipid) systems in particular is of great interest. One often talks about an ideal swelling behavior based on low-angle X-ray diffraction investigations (Fontell, 1974), which implies that, when a lamellar phase swells and incorporates water, this additional water has properties very close to those of pure water. It is possible to test this hypothesis by studies on ^2H and ^{17}O quadrupole splittings. If we assume a fixed hydration number n of amphiphilic components (and thus that additional water is free), we obtain directly from Eq. (5),

$$\Delta = |(nX_A/X_W)\chi S_b| \qquad (12)$$

Here X_A and X_W are the mole fractions of amphiphilic components and of water, respectively. It is observed for a number of cases that, in agreement with this model, a plot of Δ versus the amphiphile/water molar ratio is linear (with intercept zero) over a wide range of water concentrations (Fig. 20). Several conclusions result from these studies, such as (Persson and Lindman, 1975; Wennerström et al., 1975) the following.

(a)   There is indeed an ideal swelling where water molecules in the vicinity of the surfaces have their orientation probability distribution very

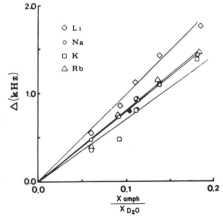

Fig. 20 Observed water ^2H quadrupole splittings for lamellar mesophase samples in the $C_7COOM–C_{10}OH–D_2O$ system as a function of the molar ratio of amphiphile to water (X_A/X_{D_2O}). The closed circles denote samples in the normal hexagonal phase of the binary system $C_7COONa–D_2O$. Temperature 20°C (Persson and Lindman, 1975)

much unchanged on increasing the water content and where water addition increases only the amount of free water (with properties quite similar to those of pure water).

(b)   There is no appreciable long-range ordering effect on the water molecules. In fact, only about one layer of water molecules (n on the order of 5 to 6) is significantly perturbed by the amphiphile aggregates. The order parameter of the free water molecules is zero within experimental uncertainty ($S_f < 10^{-3}$).

(c)   Even the water molecules in direct contact with the surfaces are very mobile and have a quite low anisotropy in their motion; generally, S_b is on the order of 10^{-2}. The order parameter of the polar part of the amphiphile molecule is typically larger by a factor of 10 or more (Tiddy, 1980, 1981).

In a number of cases, marked deviations from the simple behavior have been observed (Persson and Lindman, 1975). Then conclusion (a) does not apply. On the other hand, conclusions (b) and (c) seem to be quite general. Several reasons for the inapplicability of (a) can be proposed, but the most probable one is that there is a change in the area per polar head group resulting from a change in packing among the alkyl chains (such as interdigitation of the chains).

In principle, a combination of ^2H and ^{17}O order parameters would be expected to give additional information on how water molecules orient at surfaces. Unfortunately, however, ^2H and ^{17}O splittings give nearly linearly dependent information on the water order parameter (Halle and Wennerström, 1981b, Tricot and Niederberger, 1979). A general observation from multinuclear water splitting studies in a wide range of anisotropic systems is the invariance of the ratios of reduced splittings: The $^2H/^1H$ ratio is close to 1, and the $^{17}O/^2H$ ratio is close to 2. Several authors have considered this invariance evidence of a water orientation that is essentially independent of the molecular details of the interface. The analysis of Halle and Wennerström (1981b), however, indicates that the invariance is simply a result of the insensitivity of the splitting ratios to the shape of the water orientation probability distribution. Although this is the case for the splitting *ratios*, it is not so for the individual splittings. In fact, the influence of the chemical nature of the surface on water orientation has been amply demonstrated (Persson and Lindman, 1975; Persson and Lindman, 1977; Wennerström *et al.*, 1975). A significant influence of the nature of the polar group, charge density, counterion, solubilization, and added salt was observed.

An advantage of ^{17}O NMR is the larger quadrupole coupling constant, permitting the observation of quadrupole splittings at high water contents.

A further advantage of the ^{17}O splitting is that proton (or deuteron) exchange does not affect its magnitude. In the presence of prototropic groups in the amphiphile (—COOH, —OH, etc.), the amphiphile orientation contributes to ^2H splittings at sufficiently high deuteron exchange rates. At intermediate rates of exchange (the rate may be varied by altering the pH or the temperature) the exchange rate may be deduced from the spectral shape (Persson et al., 1973). At high exchange rates, one has the normal simple averaging, but it should be observed that the order parameter of amphiphile deuterons and water deuterons may have opposite signs (Persson and Lindman, 1975). It should also be recalled that rapid proton exchange (with amphiphilic groups or among water molecules) always causes complete elimination of the ^1H dipolar splittings (Dehl and Hoeve, 1969; Migchelsen and Berendsen, 1973) but not of the ^2H quadrupole splittings.

XI. ^{17}O-NMR Relaxation and the Hydration of Surfactant Micelles and Polyelectrolytes

The great significance of the hydration of macroscopic and microscopic interfaces and the inability of ^1H and ^2H relaxation to monitor this confidently, led Halle (1981) to exploit and develop the ^{17}O-NMR relaxation method. As mentioned previously, he used the two-step model of relaxation, which rests on the assumption that the anisotropic reorientation of hydration water (correlation time τ_c^f) occurs on a much smaller time scale than the reorientation of macromolecules or molecular aggregates (τ_c^s). The predicted values of the longitudinal and transverse relaxation rates are given, for a typical case, in Fig. 21, as a function of the resonance frequency. As can be seen, the fast motion contributes equally to both relaxation rates over the whole frequency range accessible, whereas the slow-motion contribution varies. At high frequencies, the longitudinal relaxation is determined only by the fast motion, whereas the transverse relaxation rate is always influenced by the slow motion.

Based on a simple two-site model of water molecules, the relaxation rates are expressed in terms of four parameters, τ_c^f, S, τ_c^s, and P_b. Halle et al. (Halle and Carlström, 1981; Halle and Piculell, 1982; Halle and Wennerström, 1981b; Halle et al., 1981) describe in a number of papers how it is possible to obtain these parameters for macromolecular systems from field-dependent T_1 and T_2 data. If the system can be made macroscopically anisotropic without much change in the hydration (which seems to be the case for a number of systems), the analysis is assisted a great deal by measuring quadrupole splittings of the anisotropic phase.

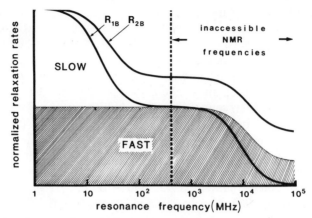

Fig. 21 Relaxation rates for hydration water according to the two-step model, with $\tau_c^f =$ 10 ps, $\tau_c^s = 5$ ns, and $S_b = 0.05$ (Halle *et al.*, 1982).

There are a number of different motions that can determine the two correlation times. τ_c^f is given either by water reorientation or by local fluctuations of some type in the macromolecule or aggregate. τ_c^s can be determined by the reorientation of the whole macromolecule or aggre-

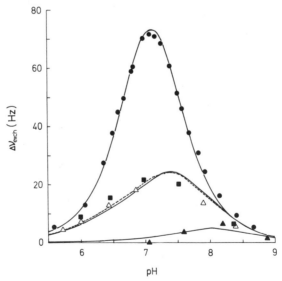

Fig. 22 Proton-exchange broadening of the ^{17}O resonance plotted against pH at 28.4°C for H_2O (●), 1.50×10^{-3} *m* polyacrylate (■, dashed curve), 10.2×10^{-3} *m* (△), and 0.585 *m* PMA (▲) (Halle and Piculell, 1982).

gate or from the limited lifetimes (τ_{ex}) of water molecules in the interfacial region. Since the aggregate tumbling often can be controlled by varying the aggregate size, it is in some cases possible to obtain τ_{ex}; thus for sufficiently large aggregates or macromolecules, τ_c^s should equal τ_{ex}. Halle and Piculell (1982) have also demonstrated an alternative possibility for obtaining τ_{ex} that involves the use of ^{17}O scalar relaxation due to proton exchange between water molecules and acidic groups on the surface (Fig. 22).

Halle *et al.* have applied these principles in experimental ^{17}O relaxation studies on a number of polyelectrolyte (Halle and Piculell, 1982), protein (Halle *et al.*, 1981), and micellar systems (Halle and Carlström, 1981). The most striking observation concerning the observed quantities that characterize hydration on the molecular level is that hydration appears to be almost the same for a wide range of systems, for example, in regard to the degree of anisotropy of hydration water, the range of perturbation exerted by a surface, the rate of reorientation of bound water molecules, and the water molecule residence time at an interface.

Halle *et al.* (1982) have summarized their conclusions as follows.

(a) In all cases, the results verify the assumptions of the two-step model. Halle and Wennerström (1981b) also demonstrate fundamental errors in the analysis in previous work on water (^1H, ^2H, and ^{17}O) relaxation in macromolecular systems as a result of, inter alia, neglect of the fast motion.

(b) Bound water molecules reorient quite rapidly, in fact at less than a factor of 10 slower than in bulk water, and the degree of anisotropy of this motion is low.

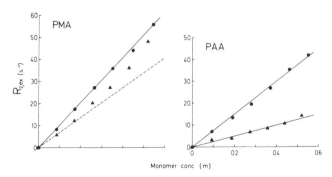

Fig. 23 Concentration dependence of the ^{17}O transverse excess relaxation rate at 34.56 MHz and 27.7°C for aqueous solutions of polyacrylate (PAA) and PMA at $\alpha = 0$ (▲) and $\alpha = 1$ (●) (Halle and Piculell, 1982).

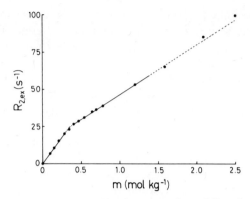

Fig. 24   Water ^{17}O transverse excess relaxation versus the molality of potassium octanoate at 27.7°C and 13.56 MHz (Halle and Carlström, 1981).

(c)   One or two layers of water molecules are perturbed appreciably, and there are no indications of long-range hydration structures.

(d)   The lifetime of a hydration water molecule is on the order of 10^{-8} s.

(e)   Charged surface groups (particularly carboxylate), as well as small counterions, induce larger structural and/or dynamic perturbations than uncharged or nonpolar groups (Fig. 23).

(f)   Less than two methylene groups in the alkyl chain are exposed to water in surfactant micelles. The much decreased water–hydrocarbon contact on micellization is evident from Fig. 24. An alkanoate in the form of micelles produces about the same effect on water ^{17}O relaxation as proprionate.

Acknowledgments

Thanks are due to Dr. Bertil Halle for critically reading the text and for suggesting revisions. Prof. Håkan Wennerström also provided useful information on the theory of counterion quadrupole relaxation in these systems. I am also grateful to Dr. Hans Gustavsson, Dr. Ali Khan, and Mr. Lennart Piculell for permission to use unpublished results.

References

Abdolall, K., Burnell, E. E., and Valic, M. I. (1977). *Chem. Phys. Lipids* **20**, 115–129.
Anderson, C. F., Record, M. T., Jr., and Hart, P. A. (1978). *Biophys. Chem.* **7**, 301–306.
Arnold, K., Loesche, A., and Gawrisch, K. (1981). *Biochim. Biophys. Acta* **645**, 143–148.

Bhuiyan, L. B., Outhwaite, C. W., and Levine, S. (1981). *Mol. Phys.* **42**, 1271–1290.
Bleam, M. L., Anderson, C. F., and Record, M. T. Jr., (1980). *Proc. Natl. Acad. Sci. USA* **77**, 3085–3089.
Bull, T. E. (1972). *J. Magn. Reson.* **8**, 344–353.
Bull, T., Forsén, S., and Turner, D. (1979). *J. Chem. Phys.* **70**, 3106–3111.
Cabane, B. (1981). *J. Phys. Orsay Fr.* **42**, 847–859.
Charvolin, J., Loewenstein, A., and Virlet, J. (1977). *J. Magn. Reson.* **26**, 529–531.
Dehl, R. E., and Hoeve, C. A. J. (1969). *J. Chem. Phys.* **50**, 3245–3251.
Dolar, D., and Peterlin, A. (1969). *J. Chem. Phys.* **50**, 3011–3015.
Edzes, H. T., Rupprecht, A., and Berendsen, H. J. C. (1972). *Biochem. Biophys. Res. Commun.* **46**, 790–794.
Eisenberg, H. (1977). *Biophys. Chem.* **7**, 3–13.
Ekwall, P. (1975). *Adv. Liq. Cryst.* **1**, 1–142.
Engström, S., and Wennerström, H. (1978). *J. Phys. Chem.* **82**, 2711–2714.
Engström, S., Jönsson, B., and Jönsson, B. (1982). *J. Magn. Reson.* (In press)
Eriksson, J. C., Johansson, Å, and Andersson, L.-O. (1966). *Acta Chem. Scand.* **20**, 2301–2304.
Fixman, M. (1979). *J. Chem. Phys.* **70**, 4995–5005.
Fontell, K. (1974). *In* "Liquid Crystals and Plastic Crystals" (G. W. Gray and P. A. Winsor, eds.), Vol. 2, pp. 80–109. Horwood, Chichester United Kingdom.
Forsén, S., and Lindman, B. (1981). *Method Biochem. Anal.* **27**, 289–486.
Grasdalen, H., and Smidsröd, O. (1981). *Macromolecules.* **14**, 1842–1845.
Guéron, M., and Weisbuch, G. (1980). *Biopolymers* **19**, 353–382.
Gunnarsson, G., Jönsson, B., and Wennerström, H. (1980). *J. Phys. Chem.* **84**, 3114–3121.
Gustavsson, H. (1982). Personal communication.
Gustavsson, H., and Lindman, B. (1975). *J. Am. Chem. Soc.* **97**, 3923–3930.
Gustavsson, H., and Lindman, B. (1978). *J. Am. Chem. Soc.* **100**, 4647–4654.
Gustavsson, H., Lindman, B., and Bull, T. E. (1978a). *J. Am. Chem. Soc.* **100**, 4655–4661.
Gustavsson, H., Siegel, G., Lindman, B., and Fransson, L.-Å. (1978b). *FEBS Lett.* **86**, 127–130.
Gustavsson, H., Siegel, G., Lindman, B., and Ehehalt, R. (1978c). *Proc. Int. Symposium on Dynamic Properties of Polyion Systems, Kyoto, 1978.*
Gustavsson, H., Siegel, G., Lindman, B., and Fransson, L.-Å. (1981). *Biochim. Biophys. Acta* **677**, 23–31.
Halle, B. (1981). "Oxygen-17 spin relaxation and molecular dynamics in aqueous systems". Ph.D. thesis, University of Lund, Lund, Sweden.
Halle, B. (1982). *Proc. of the Water in Biophysics Conference, Cambridge, 1981.* (In press)
Halle, B., and Carlström, G. (1981). *J. Phys. Chem.* **85**, 2142–2147.
Halle, B., and Piculell, L. (1982). *J. Chem. Soc. Faraday Trans. 1.* **78**, 255–271.
Halle, B., and Wennerström, H. (1981a). *J. Magn. Reson.* **44**, 81–100.
Halle, B., and Wennerström, H. (1981b). *J. Chem. Phys.* **75**, 1928–1943.
Halle, B., and Wennerström, H. (1982). Manuscript in preparation.
Halle, B., Andersson, T., Forsén, S., and Lindman, B. (1981). *J. Am. Chem. Soc.* **103**, 500–508.
Halle, B., Picullel, L., Carlström, G., Andersson, T., Wennerström, H., and Lindman, B. (1982). *Proc. of the Water in Biophysics Conference, Cambridge, 1981.* (In press)
Herwats, L., Laszlo, P., and Gerard, P. (1977). *Nouv. J. Chim.* **1**, 173–176.
Jönsson, B. (1981). "The Thermodynamics of Ionic Amphiphile-Water Systems. A Theoretical Analysis." Ph.D. Thesis, University of Lund, Lund, Sweden.
Jönsson, B., Wennerström, H., and Halle, B. (1980). *J. Phys. Chem.* **84**, 2179–2185.

Khan, A., Söderman, O., and Lindblom, G. (1980). *J. Colloid Interface Sci.* **78**, 217–224.

Khan, A., Fontell, K., and Lindblom, G. (1982a). *J. Phys. Chem.* **86**, 383–386.

Khan, A., Fontell, K., Lindblom, G., and Lindman, B. (1982b). *J. Phys. Chem.* (In press)

Kielman, H. S., and Leyte, J. C. (1973). *J. Phys. Chem.* **77**, 1593–1594.

Levij, M., de Bleijser, J., and Leyte, J. C. (1981). *Chem. Phys. Lett.* **83**, 183–191.

Levij, M., de Bleijser, J., and Leyte, J. C. (1982). *Chem. Phys. Lett.* **87**, 34–36.

Levine, S., and Outhwaite, C. W. (1978). *J. Chem. Soc. Faraday Trans. 2* **74**, 1670–1689.

Leyte, J. C. (1979). *Chem. Phys. Lett.* **66**, 417–418.

Lifson, S., and Katchalsky, A. (1954). *J. Polym. Sci.* **13**, 43–55.

Lindblom, G. (1971). *Acta Chem. Scand.* **25**, 2767–2768.

Lindblom, G. (1972). *Acta Chem. Scand.* **26**, 1745–1748.

Lindblom, G., and Lindman, B. (1973). *Mol. Cryst. Liq. Cryst.* **22**, 45–65.

Lindblom, G., Wennerström, H., and Lindman, B. (1971). *Chem. Phys. Lett.* **8**, 489–492.

Lindblom, G., Lindman, B., and Tiddy, G. J. T. (1975). *Acta Chem. Scand. A.* **29**, 876–878.

Lindblom, G., Wennerström, H., and Lindman, B. (1976). *Am. Chem. Soc. Symp. Ser.* **34**, 372–396.

Lindblom, G., Lindman, B., and Tiddy, G. J. T. (1978). *J. Am. Chem. Soc.* **100**, 2299–2303.

Lindman, B., and Brun, B. (1973). *J. Colloid Interface Sci.* **42**, 388–399.

Lindman, B., and Ekwall, P. (1968). *Mol. Cryst.* **5**, 79–93.

Lindman, B., and Forsén, S. (1976). "Chlorine, Bromine and Iodine NMR". Springer Publ., New York.

Lindman, B., and Forsén, S. (1978). *In* "NMR and the Periodic Table" (R. Harris and B. Mann, eds.), pp. 129–182, 183–194, 421–438. Academic Press, New York.

Lindman, B., and Wennerström, H. (1980). *Topics Curr. Chem.* **87**, 1–83.

Lindman, B., Lindblom, G., Wennerström, H., and Gustavsson, H. (1977). *In* "Micellization, Solubilization and Microemulsions" (K. L. Mittal, ed.), pp. 195–227. Plenum, New York.

Lindman, B., Wennerström, H., Gustavsson, H., Kamenka, N., and Brun, B. (1980a). *Pure Appl. Chem.* **52**, 1307–1315.

Lindman, B., Lindblom, G., Wennerström, H., Persson, N.-O., Gustavsson, H., and Khan, A. (1980b). *In* "Magnetic Resonance in Colloid and Interface Science (J. P. Fraissard and H. A. Resing, eds.), pp. 307–320. Reidel, New York.

Lindman, B., Puyal, M.-C., Kamenka, N., Brun, B., and Gunnarsson, G. (1982). *J. Phys. Chem.* **86**, 1702–1711.

Linse, P., and Jönsson, B. (1982). *J. Chem. Phys.* (in press)

Linse, P., Gunnarsson, G., and Jönsson, B. (1982). *J. Phys. Chem.* **86**, 413–421.

Loewenstein, A., Brenman, M., and Schwarzmann, R. (1978). *J. Phys. Chem.* **82**, 1744–1748.

Manning, G. S. (1978). *Q. Rev. Biophys.* **11**, 179–246.

Manning, G. S. (1979). *Acc. Chem. Res.* **12**, 443–449.

Meurer, B., Spegt, P., and Weill, C. (1978). *Chem. Phys. Lett.* **60**, 55–58.

Migchelsen, C., and Berendsen, H. J. C. (1973). *J. Chem. Phys.* **59**, 296–305.

Mittal, K. L. ed., (1977). "Micellization, solubilization, and microemulsions." Plenum, New York.

Mukerjee, P., and Mysels, K. J. (1971). "Critical micelle concentrations of aqueous surfactant systems" (NSRDS-NBS 36). U.S. Government Printing Office, Washington, D.C.

Niederberger, W., and Tricot, Y. (1977). *J. Magn. Reson.* **28**, 313–316.

Oosawa, F. (1971). "Polyelectrolytes". Dekker, New York.

Outhwaite, C. W. (1970). *Chem. Phys. Lett.* **7**, 636–638.

Overbeek, J. Th.C. (1976). *Pure Appl. Chem.* **46**, 91–101.

Persson, N.-O., and Lindman, B. (1975). *J. Phys. Chem.* **79**, 1410–1418.
Persson, N.-O., and Lindman, B. (1977). *Mol. Cryst. Liq. Cryst.* **38**, 327–344.
Persson, N.-O., Wennerström, H., and Lindman, B. (1973). *Acta Chem. Scand.* **27**, 1667–1672.
Persson, N.-O., Fontell, K., Lindman, B., and Tiddy, G. J. T. (1975). *J. Colloid Interface Sci.* **53**, 461–466.
Piculell, L. (1982). Personal communication.
Reeves, L. W. (1975). "Int. Review of Science, Phys. Chem. Series Two" Vol. 4, pp. 139–171. Butterworth, London.
Reimarsson, P., Parello, J., Drakenberg, T., Gustavsson, H., and Lindman, B. (1979). *FEBS Lett.* **108**, 439–442.
Reuben, J., Shporer, M., and Gabbay, E. J. (1975). *Proc. Natl. Acad. Sci. USA* **72**, 245–247.
Rose, D. M., Bleam, M. L., Record, M. T., Jr., and Bryant, R. C. (1980). *Proc. Natl. Acad. Sci.* **77**, 6289–6292.
Rosenholm, J. B., and Lindman, B. (1976). *J. Colloid Interface Sci.* **57**, 362–378.
Robb, I. D., and Smith, R. (1974). *J. Chem. Soc. Faraday Trans. 1* **70**, 287–292.
Sjöblom, E., and Henriksson, U. (1982). *J. Phys. Chem.* (In press)
Stilbs, P., and Lindman, B. (1981). *J. Phys. Chem.* **85**, 2587–2589.
Söderman, O., Engström, S., and Wennerström, H. (1980). *J. Colloid Interface Sci.* **78**, 110–117.
Tanford, C. (1973). "The Hydrophobic Effect." Wiley, New York.
Tiddy, G. J. T. (1980). *Phys. Rep.* **58**, 1–46.
Tiddy, G. J. T. (1981). *Spec. Period. Rep.* **8**, 267–294.
Tricot, Y., and Niederberger, W. (1979). *Biophys. Chem.* **9**, 195–200.
Ulmius, J., and Lindman, B. (1981). *J. Phys. Chem.* **85**, 4131–4135.
Ulmius, J., Wennerström, H., Lindblom, G., and Arvidson, G. (1977). *Biochemistry* **16**, 5742–5745.
Valleau, J. P., and Torrie, G. M. (1982). *J. Chem. Phys.* **76**, 4623–4630.
van der Klink, J. J., Zuiderweg, L. H., and Leyte, J. C. (1974). *J. Chem. Phys.* **60**, 2391–2399.
van Megen, W., and Snook, I. (1980). *J. Chem. Phys.* **73**, 4656–4662.
Wennerström, H. (1980). Personal communication.
Wennerström, H., Lindblom, G., and Lindman, B. (1974). *Chem. Scripta* **6**, 97–103.
Wennerström, H., Persson, N.-O., and Lindman, B. (1975). *Am. Chem. Soc. Symp. Ser.* **9**, 253–269.
Wennerström, H., Lindman, B., Lindblom, G., and Tiddy, G. J. T. (1979). *J. Chem. Soc. Faraday Trans 1.* **75**, 663–678.
Wennerström, H., Lindman, B., Engström, S., Söderman, O., Lindblom, G., and Tiddy, G. J. T. (1980). *In* "Magnetic Resonance in Colloid and Interface Science" (J. P. Fraissard and H. A. Resing, eds.), pp. 609–614. Reidel, New York.
Wennerström, H., Jönsson, B., and Linse, P. (1982). *J. Chem. Phys.* **76**, 4665–4670.
Westlund, P.-O., and Wennerström, H. (1982). To be published.
Westra, S. W. T., and Leyte, J. C. (1979). *Ber. Bunsenges. Phys. Chem.* **83**, 678–682.
Wong, M., Thomas, J. K., and Nowak, T. (1977). *J. Am. Chem. Soc.* **99**, 4730–4736.

9

Chlorine, Bromine, and Iodine

Björn Lindman

Physical Chemistry 1
University of Lund
Lund, Sweden

I. Introduction

The three elements Cl, Br, and I form, both from a chemical and a NMR point of view, a homogeneous group. The naturally occurring isotopes ^{35}Cl, ^{37}Cl, ^{79}Br, ^{81}Br, and ^{127}I all have sizable quadrupole moments and, as a consequence, the large majority of investigations are concerned with quadrupolar effects. For a long time, NMR investigations of all these nuclei have been abundant, although ^{35}Cl, the more receptive Cl isotope, has been studied more than the other. Applications of Cl, Br, and I NMR include studies on ions in simple aqueous or nonaqueous solutions and in complex macromolecular or surfactant systems, as well as on compounds with covalently bonded atoms. A monograph has been published that attempted comprehensive coverage of the different aspects of Cl, Br, and I NMR (Lindman and Forsén, 1976, hereafter referred to as LF). There has continued to be considerable activity in the field, with many important new results. On the other hand, there have been few basically new types of investigations and only relatively moderate progress in methodology. In view of this, the present chapter will be confined to a brief survey of work published since 1976. The underlying principles of the various studies are generally given in some detail in LF.

233

II. Halide Ions in Aqueous and Nonaqueous Solutions

The great sensitivity of both the chemical shifts and, especially, the quadrupole relaxation rates of halide ions to ion–solvent and ion–ion interactions is well established. With an appropriate theory, halogen NMR should thus be capable of providing detailed insight into a number of structural and dynamic features of electrolyte solutions, for example, ion solvation, preferential solvation in mixed solvents, and ion–ion distributions including ion pairing. The major contributions to this field of research are due to H. G. Hertz and co-workers who, for more than 20 yr, have performed experimental and theoretical investigations (cf. LF, Ch. 5). The electric field gradients at a nucleus are created by all the surrounding charges. The magnitude of the quadrupole coupling depends strongly on the symmetry of the charge distribution and may vanish partly or completely for symmetry reasons. In Hertz's electrostatic theory of ion quadrupole relaxation, the field gradients are caused by solvent molecules taken as point dipoles and by ions taken as point charges.

For aqueous solutions, novel contributions by Hertz's group concern the interactions of halide ions with various cations. Hertz and Mazitov (1981) have elaborated the electrostatic theory in the study of ion–ion distributions and on the basis of experimental cation and anion relaxation rates in combination examined the microheterogeneity of aqueous rubidium and cesium halide solutions. Helm and Hertz (1981) used ^{81}Br relaxation to study the ion–ion correlations in aqueous solutions of alkaline earth bromides. An extensive investigation of concentrated aqueous solutions of $NiCl_2$ and $NiBr_2$ used T_1 and T_2 for ^{35}Cl, ^{37}Cl, ^{79}Br, and ^{81}Br at different magnetic fields and temperatures as well as ^{35}Cl chemical shifts (Weingärtner et al., 1979). A detailed analysis taking into account the different relaxation mechanisms gave information, inter alia, on the probability of the halide ion being present in the first coordination sphere of Ni^{2+} and its lifetime there. In regard to $^{35}Cl^-$ relaxation, improved values for the infinite dilution relaxation rate as well as for the concentration dependence in alkali chloride solutions have been presented in two reports (Reimarsson et al., 1977; Holz and Weingärtner, 1977). An infinite dilution value of 25 s^{-1} at 25°C, as well as the values for $^{81}Br^-$ (1050 s^{-1}) and $^{127}I^-$ (4600 s^{-1}), are in good agreement with the predictions of Hertz's electrostatic theory (Weingärtner and Hertz, 1977).

A systematic investigation of ion quadrupole relaxation in a broad range of solvents was made by Weingärtner and Hertz (1977), and the results are shown for ^{127}I in Fig. 1. The infinite dilution values were used

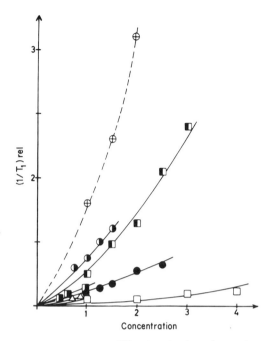

Fig. 1 The concentration dependence of ^{127}I relaxation in various solvents (25°C). Concentration scale is moles of salt per 55.5 mol of solvent. The relaxation rates are given relative to the infinite dilution values (in parentheses) for the following systems: KI–water (□, 4600 s^{-1}); RbI–N-methylformamide (●, 33,000 s^{-1}); NaI–N,N-dimethylformamide (◑, 4700 s^{-1}); $(C_2H_5)_4NI$–acetonitrile (△, 1200 s^{-1}); RbI–dimethyl sulfoxide (◧, 4000 s^{-1}); NaI–dimethyl sulfoxide (⊕); $(C_2H_5)_4NI$–dimethyl sulfoxide (▣) (Weingärtner and Hertz, 1977).

to provide information on ion solvation, for example, solvent molecule distribution around the halide ion. Halide ion solvation is found to be weak in formamide, acetone, acetonitrile, and dimethyl sulfoxide, whereas methanol, ethanol, and formic acid are characterized by much more effective relaxation and thus stronger solvation. These differences were discussed in terms of hydrogen bonding in the solvation process for the protic solvents. The ion–ion effects gave a complex pattern explained in terms of local structural effects and of cross-correlation between ions and solvent molecules.

Preferential solvation in various mixed solvents has been examined by Holz *et al.* (Holz, 1978; Holz *et al.*, 1977, 1978). For methanol–water mixtures, an analysis of the dependence of the extrapolated infinite dilution ^{35}Cl and ^{81}Br relaxation rates on solvent composition suggests that both Cl$^-$ and Br$^-$ are preferentially solvated by methanol; for I$^-$, approximately nonpreferential solvation appears to apply (Holz *et al.*, 1977). For

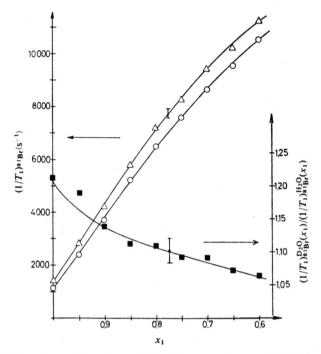

Fig. 2 ^{81}Br relaxation (left-hand scale) in $CH_3OH–H_2O$ (○) and $CH_3OH–D_2O$ (△) (1 m NaBr, concentration scale as in Fig. 1) as a function of the mole fraction of water in the solvent (x_1). ■, Ratio between the two relaxation rates (right-hand scale) (Holz, 1978).

some water–amide mixtures, an analysis in terms of selective solvation effects is difficult because of the effects on anion relaxation connected with hydrophobic hydration (Holz *et al.*, 1978). Holz (1978) used the water hydrogen–deuterium isotope effect on ion relaxation to deduce that Br^- is preferentially solvated by methanol but not by formamide mixed with water. His method, illustrated in Fig. 2, is interesting but is limited in precision by the fact that the maximal isotope effect is only ~20% at 25°C.

An important effect of hydrophobic groups on the quadrupole relaxation of aqueous halide ions has been observed for several systems and has been attributed to large field gradients arising from assymmetric solvation (LF, Ch. 5). Yudasaka *et al.* performed ^{35}Cl studies on several systems (Kato *et al.*, 1982; Yudasaka *et al.*, 1981). Systems investigated included solutions of tetraethylammonium chloride and Cl^- ions in mixtures of water and other solvents. As an example (Fig. 3a), in mixtures of water and dimethyl sulfoxide, the relaxation rates at intermediate solvent compositions are higher than those of either pure solvent by a factor of more

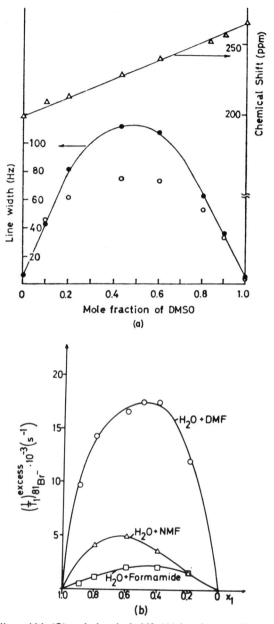

Fig. 3 (a) $^{35}Cl^-$ line width (●) and chemical shift (△) in mixtures of water and dimethyl sulfoxide. Also known is the viscosity-corrected line width (○). (Yudasaka *et al.*, 1982) (b) Excess (after correction for correlation times) longitudinal ^{81}Br relaxation rates in mixtures of water (mole fraction x_1) and formamide (□), N-methylformamide (△), and N, N-dimethylformamide (○) (Holz *et al.*, 1978).

than 10; the ^{35}Cl chemical shift, on the other hand, varies linearly with solvent composition. The ^{81}Br relaxation studies of Holz *et al.* (1978) on water–amide mixtures illustrate well the role of hydrophobic effects (Fig. 3b).

For concentrated aqueous solutions of hydrobromic acid, both ^{81}Br quadrupole relaxation and chemical shifts gave accurate information on acid dissociation (Soffer and Marcus, 1982), but $AlCl_3$ in acetonitrile was difficult to study (Beattie *et al.*, 1979).

In regard to the theory of ion quadrupole relaxation, in addition to Hertz's work, the electrostatic continuum approach of Hynes and Wolynes (1981) has been reported. A purely molecular approach to ion quadrupole relaxation has also been described (Engström and Jönsson, 1981; Engström *et al.*, 1982), where the electric field gradient fluctuation for *inter alia* infinitely dilute aqueous Cl^- ions was calculated by means of Monte Carlo simulations. The potential energy functions and the field gradient expressions were obtained from accurate (ab initio) quantum chemical calculations. The results have been compared with Hertz's electrostatic theory, and both qualitative and quantitative differences found. For Cl^-, both water translation and rotation give sizable fluctuations in the field gradient, and it appears that, in general, motions on two different time scales contribute to the relaxation. Engström *et al.* (1982) also discuss the magnitude of the polarization factor used in the electrostatic theory and suggest that molecular dynamics simulations would give a much deeper insight into ion quadrupole relaxation.

III. The ClO_4^- Ion

Because of the tetrahedral symmetry, ^{35}Cl quadrupole relaxation of the perchlorate ion is slow, and thus its resonance signal is easy to study. Reimarsson *et al.* (1977) determined the infinite dilution relaxation rate of the aqueous ion to be $3.6 \ s^{-1}$ at 25°C and discussed the relaxation mechanism which is much more complex than for monatomic ions. It has been found that both distortions from the tetrahedral symmetry and an electric field-induced field gradient must be taken into account. The concentration dependence of the $^{35}ClO_4^-$ relaxation in aqueous solutions is very sensitive to the cation, the interaction following the order $Fe^{3+} > Ba^{2+} > Ca^{2+} > Mg^{2+} > Li^+ > Na^+ > NH_4^+ > H^+$ (Reimarsson and Lindman, 1977). The ^{35}Cl and ^{37}Cl relaxation in aqueous solutions of $Mn(ClO_4)_2$ is dominated by interaction with the unpaired electron spin of Mn^{2+} (Reimarsson, 1980). ClO_4^- was shown to form an inner-sphere complex with Mn^{2+}, and the relaxation data gave values for the rotational correlation time and the lifetime of ClO_4^- in the complex. The ^{35}Cl relaxation of aqueous ClO_4^- has also been investigated to obtain the ion–ion and ion–solvent distributions with alkali or alkaline earth ions as cations (Contreras and Hertz, 1978a,b; Helm and Hertz, 1981) and to study the interaction of ClO_4^- with NpO_2^{2+} in water–acetone mixtures (Shcherbakov and Jorga, 1977).

Greenberg and Popov (1976) inferred a quite weak interaction of ClO_4^- with Na^+ in several solvents from ^{35}Cl relaxation, and Grandjean and Laszlo (1977) demonstrated an effect of sorbose on the ion–ion contribution to relaxation in pyridine. ^{35}Cl relaxation demonstrates a strong interaction between ClO_4^- and Ca^{2+} in hexamethylphosphoramide, whereas the interaction with Mg^{2+} and Ba^{2+} is weak (Echegoyen et al., 1981).

The chemically related IO_4^- ion gives narrow ^{127}I resonance signals in aqueous solution (Reimarsson et al., 1977; Segal and Vyas, 1980). The infinite dilution relaxation rate is $\sim 5 \times 10^3$ s^{-1} and has been used to estimate the Sternheimer shielding (Reimarsson et al., 1977).

IV. Surfactant Systems

A major application of halogen NMR involves the study of ionic interactions in colloidal systems, mainly aqueous solutions of amphiphiles, proteins, and polyelectrolytes. The quadrupole relaxation rate is very sensitive to these interactions, in most cases mainly a result of contributions from slow motional processes, but the chemical shift and the quadrupole splitting are also very useful, especially for surfactant systems. An introduction to such studies is given in Chapter 8 and reference is made to a more comprehensive account (Forsén and Lindman, 1981) and to introductions to the field (Forsén, 1978; Forsén and Lindman, 1978; Lindman, 1978; Lindman et al., 1977). Recent studies in the field of amphiphilic systems have focused on phase structure, the ion condensation model of counterion binding, and counterion specificity and hydration.

For hexadecyltrimethylammonium halides ($C_{16}TAX$), ^{35}Cl and ^{81}Br relaxation demonstrate at room temperature a pronounced change in micellar shape from spheres to rods for $C_{16}TABr$ but not for $C_{16}TACl$ (Ulmius et al., 1978; Fabre et al., 1980). For mixed solutions of the two surfactants, ^{81}Br relaxation corresponds to that of a rod micelle solution, and ^{35}Cl relaxation to that of a solution containing only spherical micelles. This behavior can be rationalized in terms of a preferential binding of Br^- ions to both spherical and rodlike micelles and of a greater tendency for Br^- to induce micelle growth. At low temperatures, $C_{16}TACl$ also produces rod micelles. In solutions of $C_{12}TACl$ micelles, there is a difference in the ^{35}Cl chemical shift between H_2O and D_2O solutions of 3.8–3.9 ppm (concentration 1–2 m) (Gustavsson and Lindman, 1978). This is not much below the value for infinite dilution (4.4 ppm), suggesting that bound Cl^- counterions remain hydrated. For Br^-, the situation may be different.

As described in Chapter 8, counterion quadrupole splittings (Lindblom et al., 1976) normally give direct information on variations in the degree of counterion binding. ^{35}Cl quadrupole splittings (see Fig. 14 in Chapter 8) have contributed to establishing an ion condensation-type behavior characterized by a near insensitivity of counterion binding to temperature, concentration, and electrolyte addition (Lindman et al., 1980; Wennerström et al., 1979; Wennerström et al., 1980). For the hexagonal liquid crystalline phase of the system $C_{16}TACl–C_{16}TABr–H_2O$, ^{35}Cl splittings demonstrate a preferential binding of Br^- ions in competition experiments (Fabre et al., 1980). Counterion binding in the "nematic lyotropic" mesophase of the system $C_{10}NH_3Cl–NH_4Cl–H_2O$ was investigated by Fujiwara and Reeves (1976) using ^{35}Cl splittings. In a nematic lyotropic phase of disklike micelles, the Br^- binding varies, according to ^{81}Br splittings, with the ratio between cationic and anionic surfactants, as expected (Lee et al., 1980).

V. Macromolecular Systems

Halogen quadrupole relaxation has become a standard method in the study of anion binding to proteins, but it has been used much less for other macromolecular systems (LF, Ch. 8; Forsén and Lindman, 1981). The most important use of the technique is as a titration indicator where competition experiments, variable halide concentration studies, pH variations, chemical modifications, etc., are used to obtain, rather conveniently, information on ion binding affinity, sites of anion binding, cooperativity, the roles of metal ions in binding, etc. Additional information, both structural and dynamic, on the molecular aspects of binding can be obtained from a closer analysis of the relaxation rates. This requires access to variable-field T_1 and T_2 data and preferably to the nonexponentiality of relaxation. It is well established that, for ions bound to macromolecules, motions on different times scales contribute to relaxation. As described in Chapter 8, it appears that Wennerström and Halle's two-step model of relaxation (Halle and Wennerström, 1981; Wennerström et al., 1974), with a rapid local motion and a slow overall motion, is an appropriate approximation for the analysis of ion relaxation data in macromolecular systems. Applications of this theory to ion binding studies are still in an initial stage, but a direct implication is that generally an important fraction of the total interaction is modulated by rapid local motions. Deduced quadrupole coupling constants, i.e., those modulated by slow motions, should therefore be interpreted with care in terms of mode of binding, symmetry, etc.

^{35}Cl investigation of anion binding to proteins is the most studied area, and proteins recently considered include lactate dehydrogenase (Bull *et al.*, 1976; Reimarsson *et al.*, 1978), malate dehydrogenase (Reimarsson *et al.*, 1978), different hemoglobins (Chiancone *et al.*, 1981; Norne *et al.*, 1978), erythrocruorine (Chiancone *et al.*, 1976), an anion transport protein of erythrocytes (Shami *et al.*, 1977), chloroperoxidase (Krejcarek *et al.*, 1976), superoxide dismutase (Fee and Ward, 1976), alkaline phosphatase (Norne *et al.*, 1979), liver alcohol dehydrogenase (Andersson *et al.*, 1978, 1979a), human plasma albumin (Halle and Lindman, 1978; Bull *et al.*, 1978a), cytochrome c (Andersson *et al.*, 1979b), liver dihydrofolate reductase (Subramanian *et al.*, 1981), carbonic anhydrase (Ward, 1980), halophilic ferredoxin (Reimarsson *et al.*, 1980), and human plasma transferrin (Linse *et al.*, 1981). A review of the field has been presented by Stephens and Bryant (1976), and a study on α-chymotrypsin with ^{79}Br and ^{81}Br NMR has been reported by Garnett *et al.* (1976). It may also be mentioned that ^{35}Cl NMR has been used to investigate ion binding to nucleosides in dimethyl sulfoxide (Plaush and Sharp, 1976), to amino acids in water (Jönsson and Lindman, 1977), to aqueous zinc–nucleotide triphosphate complexes (Happe and Ward, 1979), and to bacterial surfaces (Lyon *et al.*, 1976).

As a brief example, let us take the studies of internal motion at anion-binding sites in proteins (Bull, 1978; Bull *et al.*, 1978b) and on coenzyme binding to horse liver alcohol dehydrogenase (Andersson *et al.*, 1979a). Based on a simple model of the internal motion corresponding to a rigid rod in a cone, ^{35}Cl relaxation data for several proteins were analyzed. If the quadrupole coupling constant is known, it is possible to deduce the extent and speed of the internal motions. With the quadrupole coupling constant assumed to be 3.6 MHz, the cone angle and the internal correlation time were obtained. An interesting finding was that the latter was about 10^{-9} s for all cases studied (binding to nonmetal sites).

Horse liver alcohol dehydrogenase, which is a zinc metalloenzyme, shows no Cl$^-$ binding to metal ions, whereas Cl$^-$ is extensively eliminated by a coenzyme (NADH) (Andersson *et al.*, 1979a). The anion binding is complex, with at least two anion-binding sites within the coenzyme binding domain. A mapping of the anion binding was obtained from competition experiments with coenzyme fragments and coenzyme competitive inhibitors, as illustrated in Fig. 4.

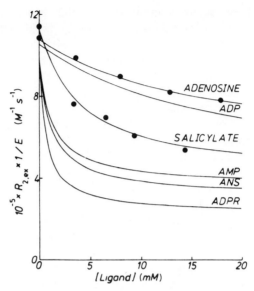

Fig. 4 ^{35}Cl relaxation in solutions of horse liver alcohol dehydrogenase and coenzyme fragments or inhibitors. The excess transverse relaxation rate (relative to that of a protein-free solution) divided by the enzyme concentration is given as a function of the concentration of adenosine, adenosine monophosphate (AMP), adenosine diphosphate (ADP), adenosine diphosphoribose (ADPR), salicylate, and 8-anilino-1-naphthalene sulfonate (ANS). The Cl$^-$ concentration was 0.5 M, and the enzyme concentration 60–100 μM (Andersson *et al.*, 1979a).

In regard to polycation systems, the very large effects on ^{37}Cl relaxation in solutions of polydimethylaminoethyl methacrylate may be mentioned (Gustavsson and Lindman, 1976). To produce manageable effects, it was necessary to have a large excess of Cl$^-$ ions. This study included investigations of ampholytic copolymers and terpolymers and somewhat unexpectedly measured the small effect of a polyanion, polymethacrylate, on ^{37}Cl$^-$ relaxation; the effect is reduced when cation condensation starts. Grasdalen and Smidsröd (1981) recently reported an interesting observation using ^{127}I to study a negatively charged polysaccharide, κ-carrageenan. A particularly strong temperature-dependent (Fig. 5) ^{127}I$^-$ binding was demonstrated.

VI. Covalently Bonded Halogens

The Cl relaxation of simple liquids has been one of the major sources of information on molecular reorientation in liquids (LF, Ch. 2). For Br and

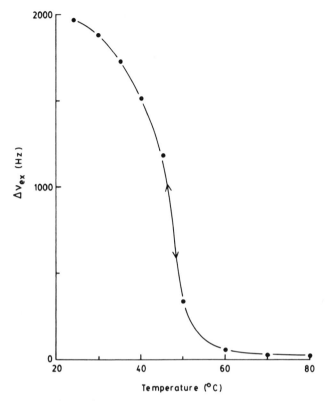

Fig. 5 Temperature dependence of the excess ^{127}I line width for solutions of tetramethyl-ammonium κ-carrageenate. The concentration of polysaccharide was 0.15% (w/v) and of I$^-$ 0.15 M (Grasdalen and Smidsröd, 1981).

I, relaxation is so rapid that, except for special cases, relaxation times are generally obtained indirectly in studies on another nucleus relaxed by scalar interactions. Spin–spin couplings are also obtained in this way (LF, Ch. 4). It is only for Cl that resonance lines are narrow enough to permit studies on shielding effects (LF, Ch. 3).

In a series of extensive ^{35}Cl investigations, Barlos et al. (Barlos et al., 1977; Barlos et al., 1978; Barlos et al., 1980) have provided much novel information on the shielding of covalently bonded Cl. Compounds investigated included chloroboranes, chlorosilanes, chloroarsanes, alkyl and alkenyl chlorides, and carboxylic acid chlorides. The results were discussed in terms of types of bonding and charge density, but theoretical rationalization is difficult. That medium effects on shielding and not only intramolecular effects are important was demonstrated in a ^{35}Cl and ^{37}Cl study on TiCl$_4$ in different solvents (Forsén et al., 1976). This work in-

cluded a demonstration, using Cl relaxation for several compounds, that
molecular reorientation was strongly dependent on the medium. Other
^{35}Cl studies have involved the T_2 of neat p-dichlorobenzene and $TiCl_4$
(Rizaev, 1980) and of chloroaluminate melts (Anders and Plambeck,
1978).

Relaxation studies on other nuclei to which a halogen is spin-coupled
have in a number of cases provided information on quadrupole relaxation
rates and spin–spin coupling constants. This includes ^{119}Sn studies on
$SnCl_3I$ and SnI_3Cl (Sharp and Tolan, 1976), ^{1}H, ^{2}H, and ^{13}C studies on
CH_3Br (Lassigne and Wells, 1977), ^{13}C studies on a large number of
bromo compounds (bromomethanes, bromobenzenes, etc.) (Hayashi *et
al.*, 1980; Yamamoto and Yanagisawa, 1977; Yanagisawa and Yamamoto,
1979), and ^{27}Al studies on Al^{3+} complexes with Cl^- and Br^- (Tarasov *et
al.*, 1978). In addition, ^{35}Cl spin couplings have been determined to ^{17}O in
ClO_3F, ClO_4^-, $ClOF_3$, ClO_2F, and ClO_2AsF_6 with ^{17}O NMR (Virlet and
Tantot, 1976) and to ^{17}O in ClO_4^- with ^{17}O NMR (Lutz *et al.*, 1976).

Two studies pioneering the field of Cl relaxation in the gas phase have
given interesting information on gas phase molecular motion. Lee and
Cornwell (1976) determined the T_2 of ^{35}Cl from line widths for gaseous
HCl in the pressure range 1–35 atm and also chemical shifts, providing an
absolute reference for Cl chemical shifts. Gillen *et al.* (1978) performed
^{19}F and ^{35}Cl relaxation studies on gaseous as well as liquid and plastic
crystalline ClF. These data provided the ^{19}F–^{35}Cl spin coupling constant
as well as rather complete information on the molecular motion (reorientational and angular velocity correlation times), permitting a comparison
with current theories.

VII. Conclusions

The applications of Cl, Br, and I NMR are in many respects parallel to
those of other quadrupolar nuclei. For halide ions, the theoretical principles and problems studied are essentially the same as for alkali ions. In
addition, Cl NMR has found important applications in studies on the
covalent state; here the relaxation times of Br and I are generally very
short (below 10^{-6} s).

The relaxation times of Cl^-, Br^-, I^-, and ClO_4^- ions have provided
insight into ion–solvent and ion–ion interactions in a wide range of systems. Important for future developments is further work on electrostatic
models of ion quadrupolar relaxation, as well as Monte Carlo and molecular dynamics simulations of this problem. In particular, ^{35}Cl NMR has

become a standard method for studying ionic interactions in protein and amphiphile systems, whereas interest in polyelectrolyte and complex macromolecular systems has been remarkably low. This situation is now changing. For macromolecular systems, the analysis of multifield T_1 and T_2 data can be expected to provide detailed insight into various motion processes.

References

Anders, U., and Plambeck, J. A. (1978). *J. Inorg. Nucl. Chem.* **40**, 387–338.

Andersson, I., Katzberg, D., Lindman, B., and Zeppezauer, M. (1978). *Dev. Halophilic Microorg.* **1**, pp. 33–40.

Andersson, I., Zeppezauer, M., Bull, T., Einarsson, R., Norne, J. E., and Lindman, B. (1979a). *Biochem.* **18**, 3407–3413.

Andersson, T., Thulin, E., and Forsén, S. (1979b). *Biochem.* **18**, 2487–2493.

Barlos, K., Kroner, J., Nöth, H., and Wrackmeyer, B. (1977). *Chem. Ber.* **110**, 2774–2782.

Barlos, K., Kroner, J., Nöth, H., and Wrackmeyer, B. (1978). *Chem. Ber.* **111**, 1833–1838.

Barlos, K., Kroner, J., Nöth, H., and Wrackmeyer, B. (1980). *Chem. Ber.* **113**, 3716–3723.

Beattie, I. R., Jones, P. J., Howard, J. A. K., Smart, L. E., Gilmore, C. J., and Akitt, J. W. (1979). *J. Chem. Soc. Dalton Trans.* **3**, 528–535.

Bull, T. E. (1978). *J. Magn. Reson.* **31**, 453–458.

Bull, T. E., Lindman, B., and Reimarsson, P. (1976). *Arch. Biochem. Biophys.* **176**, 389–391.

Bull, T. E., Halle, B., and Lindman, B. (1978a). *FEBS Lett.* **86**, 25–28.

Bull, T. E., Norne, J.-E., Reimarsson, P., and Lindman, B. (1978b). *J. Am. Chem. Soc.* **100**, 4643–4647.

Chiancone, E., Bull, T. E., Norne, J.-E., Forsén, S., and Antonini, E. (1976). *J. Mol. Biol.* **107**, 25–34.

Chiancone, E., Norne, J.-E., and Forsén, S. (1981). *Methods Enzymol.* **76**, 552–559.

Contreras, M., and Hertz, H. G. (1978a). *Faraday Disc.* **64**, 33–47.

Contreras, M., and Hertz, H. G. (1978b). *J. Solution Chem.* **7**, 99–137.

Echegoyen, L., Nieves, I., Thompson, J., Rosa, F., and Concepcion, R. (1981). *J. Phys. Chem.* **85**, 3697–3699.

Engström, S., and Jönsson, B. (1981). *Mol. Phys.* **43**, 1235–1253.

Engström, S., Jönsson, B., and Jönsson, B. (1982). *J. Magn. Reson.* (In press)

Fabre, H., Kamenka, N., Khan, A., Lindblom, G., Lindman, B., and Tiddy, G. J. T. (1980). *J. Phys. Chem.* **84**, 3428–3433.

Fee, J. A., and Ward, R. L. (1976). *Biochem. Biophys. Res. Commun.* **71**, 427–437.

Forsén, S. (1978). *In* "Proc. Eur. Conf. NMR Macromol" (E. Conti, ed.), pp. 243–296. Lerici, Sardinia.

Forsén, S., and Lindman, B. (1978). *Chem. Br.* **14**, 29–35.

Forsén, S., and Lindman, B. (1981). *Methods Biochem. Anal.* **27**, 289–486.

Forsén, S., Gustavsson, H., Lindman, B., and Persson, N. O. (1976). *J. Magn. Reson.* **23**, 515–518.

Fujiwara, F. Y., and Reeves, L. W. (1976). *J. Am. Chem. Soc.* **98**, 6790–6798.

Garnett, M. W., Halstead, T. K., and Hoare, D. G. (1976). *Eur. J. Biochem.* **66**, 85–93.

Gillen, K. T., Douglass, D. C., and Griffiths, J. E. (1978). *J. Chem. Phys.* **69**, 461–467.

Grandjean, J., and Laszlo, P. (1977). *Helv. Chim. Acta* **60**, 259–261.

Grasdalen, H., and Smidsröd, O. (1981). *Macromolecules.* **14**, 1842–1845.

Greenberg, M. S., and Popov, A. I. (1976). *J. Solution Chem.* **5**, 653–665.

Gustavsson, H., and Lindman, B. (1976). In "Colloid Interface Sci., Proc. Int. Conf., 50th" (M. Kerker, ed.), pp. 339–355, Vol. 2. Academic Press, New York.

Gustavsson, H., and Lindman, B. (1978). *J. Am. Chem. Soc.* **100**, 4647–4654.

Halle, B., and Lindman, B. (1978). *Biochemistry* **17**, 3774–3781.

Halle, B., and Wennerström, H. (1981). *J. Chem. Phys.* **75**, 1928–1943.

Happe, J. A., and Ward, R. L. (1979). *J. Phys. Chem.* **83**, 3457–3462.

Hayashi, S., Hayamizu, K., and Yamamoto, O. (1980). *J. Magn. Reson.* **37**, 17–29.

Helm, L., and Hertz, H. G. (1981). *Z. Phys. Chem. Weisbaden* **127**, 23–44.

Hertz, H. G., and Mazitov, R. K. (1981). *Ber. Bunsenges. Phys. Chem.* **85**, 1103–1112.

Holz, M. (1978). *J. Chem. Soc. Faraday Trans. 1* **74**, 644–656.

Holz, M., and Weingärtner, H. (1977). *J. Magn. Reson.* **27**, 153–155.

Holz, M., Weingärtner, H., and Hertz, H. G. (1977). *J. Chem. Soc. Faraday Trans. 1* **73**, 71–83.

Holz, M., Weingärtner, H., and Hertz, H. G. (1978). *J. Solution Chem.* **7**, 705–720.

Hynes, J. T., and Wolynes, P. G. (1981). *J. Chem. Phys.* **75**, 395–401.

Jönsson, B., and Lindman, B. (1977). *FEBS Lett.* **78**, 67–71.

Kato, T., Yudasaka, M., and Fujiyama, T. (1982). *Bull. Chem. Soc. Jpn.* **55**, 1284–1289.

Krejcarek, G. E., Bryant, R. G., Smith, R. J., and Hayer, L. P. (1976). *Biochemistry* **15**, 2508–2511.

Lassigne, C. R., and Wells, E. J. (1977). *J. Magn. Reson.* **27**, 215–236.

Lee, C. Y., and Cornwell, C. D. (1976). *In* "Magnetic Resonance and Related Phenomena," pp. 261–264. Groupement Ampere, Heidelberg.

Lee, Y., Reeves, L. W., and Tracey, A. S. (1980). *Can. J. Chem.* **58**, 110–123.

Lindblom, G., Wennerström, H., and Lindman, B. (1976). *Am. Chem. Soc. Symp. Ser.* **34**, 372–396.

Lindman, B. (1978). *J. Magn. Reson.* **32**, 39–47.

Lindman, B., and Forsén, S. (1976). "Chlorine, Bromine and Iodine NMR." Springer, Heidelberg.

Lindman, B., Lindblom, G., Wennerström, H., and Gustavsson, H. (1977). *In* "Micellization, Solubilization and Microemulsions" (K. L. Mittal, ed.), pp. 195–227. Plenum, New York.

Lindman, B., Lindblom, G., Wennerström, H., Persson, N.-O., Gustavsson, H., and Khan, A. (1980). *In* "Magnetic Resonance in Colloid and Interface Science" (J. P. Fraissard and H. A. Resing, eds.), pp. 307–320. Reidel, New York.

Linse, P., Einarsson, R., and Lindman, B. (1981). *Chem. Scripta* **18**, 4–6.

Lutz, O., Nepple, W., and Nolle, A. (1976). *Z. Naturforsch.* **31a**, 1046–1050.

Lyon, R. C., Magnuson, N. S., and Magnuson, J. A. (1976). *In* "Extreme Environments, Mechanisms of Microbial Adaption," pp. 305–320. Academic Press, New York.

Norne, J.-E., Chiancone, E., Forsén, S., Antonini, E., and Wyman, J. (1978). *FEBS Lett.* **94**, 410–412.

Norne, J.-E., Szajn, H., Csopak, H., Reimarsson, P., and Lindman, B. (1979). *Arch. Biochem. Biophys.* **96**, 552–556.

Plaush, A. C., and Sharp, R. R. (1976). *J. Am. Chem. Soc.* **98**, 7973–7980.

Reimarsson, P. (1980). *J. Magn. Reson.* **38**, 245–252.

Reimarsson, P., and Lindman, B. (1977). *Inorg. Nucl. Chem. Lett.* **13**, 449–453.

Reimarsson, P., Wennerström, H., Engström, S., and Lindman, B. (1977). *J. Phys. Chem.* **81**, 789–792.

Reimarsson, P., Bull, T., Norne, J.-E., and Lindman, B. (1978). *In* "Energetics and Structure of Halophilic Microorganisms" (S. R. Caplan and M. Ginsberg, eds.), pp. 41–48. Elsevier, New York.

Reimarsson, P., Lindman, B., and Werber, M. M. (1980). *Arch. Biochem. Biophys.* **202**, 664–666.

Rizaev, V. R. (1980). *Ukr. Fiz. Zh.* **5**, 860–862.

Segal, S. L., and Vyas, H. M. (1980). *J. Chem. Phys.* **72**, 1406–1407.

Shami, Y., Carver, J., Ship, S., and Rothstein, A. (1977). *Biochem. Biophys. Res. Comm.* **76**, 429–436.

Sharp, R. R., and Tolan, J. W. (1976). *J. Chem. Phys.* **65**, 522–530.

Shcherbakov, V. A., and Jorga, E. V. (1977). *INIS Atomindex* **8**(14); *Chem. Abstr.* **87**, 142076.

Soffer, N., and Marcus, Y. (1982). *Ber. Bunsenges. Phys. Chem.* **86**, 72–73.

Stephens, R., and Bryant, R. G. (1976). *Mol. Cell. Biochem.* **13**, 101–112.

Subramanian, S., Shindo, H., and Kaufman, B. T. (1981). *Biochem.* **20**, 3226–3230.

Tarasov, V. P., Privalov, V. I., and Buslaiv, Y. A. (1978). *J. Struct. Chem. USSR* **19**, 866–871.

Ulmius, J., Lindman, B., Lindblom, G., and Drakenberg, T. (1978). *J. Coll. Interface Sci.* **65**, 88–97.

Virlet, J., and Tantot, G. (1976). *Chem. Phys. Lett.* **44**, 296–299.

Ward, R. L. (1980). *In* "Biophys. Physiol. Carbon Dioxide, Symp." (C. Bauer, G. Gros, and H. Bartels, eds.), pp. 262–272. Springer, Berlin.

Weingärtner, H., and Hertz, H. G. (1977). *Ber. Bunsenges. Phys. Chem.* **81**, 1204–1221.

Weingärtner, H., Müller, C., and Hertz, H. G. (1979). *J. Chem. Soc. Faraday Trans. 1* **75**, 2712–2734.

Wennerström, H., Lindblom, G., and Lindman, B. (1974). *Chem. Scripta* **6**, 97–103.

Wennerström, H., Lindman, B., Lindblom, G., and Tiddy, G. J. T. (1979). *J. Chem. Soc. Faraday Trans. 1* **75**, 663–668.

Wennerström, H., Lindman, B., Engström, S., Söderman, O., Lindblom, G., and Tiddy, G. J. T. (1980). *In* "Magnetic Resonance in Colloid and Interface Science" (J. P. Fraissard and H. A. Resing, eds.), pp. 609–614. Reidel, New York.

Yamamoto, O., and Yanagisawa, M. (1977). *J. Chem. Phys.* **67**, 3803–3807.

Yanagisawa, M., and Yamamoto, O. (1979). *Bull. Chem. Soc. Jpn.* **52**, 2147–2148.

Yudasaka, M., Sugawara, T., Iwamura, H., and Fujiyama, T. (1981). *Bull. Chem. Soc. Jpn.* **54**, 1933–1938.

Yudasaka, M., Sugawara, T., Iwamura, H., and Fujiyama, T. (1982). *Bull. Chem. Soc. Jpn.* **55**, 311–312.

10 Antibiotic Ionophores

Hadassa Degani

Isotope Department
The Weizmann Institute of Science
Rehovot, Israel

The application of NMR to the study of antibiotic ionophores is an example of the ability of this type of spectroscopy to characterize the structure, conformation, and interactions of molecules in solution and in more complex systems such as membranes.

In Section I the discovery and function of these antibiotics in biological systems will be described briefly, together with investigations of their chemical and physical properties. More detailed information related to this introduction can be found in several reviews (McLaughlin and Eisenberg, 1975; Ovchinnikov *et al.,* 1974; Pressman, 1976). In Sections II–IV various NMR methodologies and their specific application to the study of ionophores and their ionic complexes will be described.

Section II is devoted to the characterization by NMR of the structure and conformation in solution of free and ion-complexed antibiotics. Instead of briefly describing structural studies related to all known ionophores, we have focused on one example, the most common ionophore—valinomycin.

Section III summarizes NMR methods and data related to the solution chemistry of ion–ionophore interactions, including equilibrium and kinetic studies.

Section IV describes the extension of NMR studies, similar to those performed in solution (Sections II and III), to membrane systems. This relatively new and challenging area is particularly emphasized because of its contribution to our understanding of the mechanism of action of ionophores in physiological processes. Although this chapter is confined to natural ionophores, it is important to mention that many NMR studies have also been performed with synthetic ionophores such as crown ethers and cryptates.

I. Introduction

A. Discovery and Classification

During the 1960s, in the course of a search for new antibiotics, a novel class of microbial metabolites later termed ionophores (Pressman *et al.*, 1967) or membrane-active complexones (Ovchinnikov *et al.*, 1974) was discovered. The antibiotic valinomycin, originally isolated from a *Streptomyces* fermentation (Brockmann and Schmidt-Kastner, 1955), was the first observed to mediate ion transport by stimulating changes in energy-linked ion gradients in mitochondria (Moore and Pressman, 1964). Further investigations showed that valinomycin as well as other antibiotics could induce specific ion permeability across membranes by carrying the ions selectively through the lipid barrier as lipid-soluble complexes. Most ionophores were found to act as mobile carriers, but others, such as gramicidin A and alamethecin, mediated ion transport via the formation of ion-conducting channels across membranes (Gordon and Haydon, 1972; Hladky and Haydon, 1970).

Mobile carriers are divided into two main groups: (a) neutral ionophores (first called valinomycin-type ionophores) which form positively charged complexes with cations, thereby acting as electrogenic or electrophoretic carriers, and (b) carboxylic ionophores (initially called nigericin-type ionophores) which usually form neutral ionic complexes that can cross membranes in an electrically passive fashion and are able to induce ion–proton exchange diffusion. A list of the most common antibiotic ionophores, including their structure and ionic specificity, is given in Table I.

B. Biological Application

Ionophores have been employed extensively in studying energy-coupled ionic fluxes as well as passive ion diffusion in many systems including mitochondria, chloroplasts, chromatophores, bacterial cells, and more recently muscles and nerve cells. Many of these studies contributed to the establishment and understanding of Mitchell's chemiosmotic theory. This theory in turn helped demonstrate that the mediation of transport by neutral ionophores promotes dissipation of the energy-dependent membrane potential, whereas carboxylic ionophores cause the collapse of pH gradients.

In recent years the range of application of ionophores has been considerably widened by discovery of the ionophores A-23187 (Reed and Lardy, 1972) and X-537A (also called lasalocid) (Berger *et al.*, 1951; Westley *et al.*, 1970), which interact specifically with divalent ions and can therefore serve as tools in investigations of Ca^{2+}- and Mg^{2+}-regulated processes.

C. Chemical and Physical Properties

The mechanism of carrier-mediated transport involves the following steps: (a) formation of a carrier–metal complex either in solution or at the interface of the membrane and the solution, (b) transport of the complex through the membrane, (c) dissociation of the metal from the carrier and its release either into the solution or at the membrane interface, and (d) transport of the carrier back through the membrane (either free or protonated).

One can study each step separately by characterizing the structure as well as the chemical and physical properties of the ionophore and its complexes in a homogeneous solution and in a model membrane system.

The crystal structure of many ionophores, both free and bound to ions, has been determined by X-ray crystallography. In general these studies indicate that all mobile carrier ionophores have a cyclic or quasi-cyclic structure stabilized by hydrogen bonds with electron-donating groups (oxygen or nitrogen atoms) capable of complexing ions directed toward the interior part of the complex. They also have a highly hydrophobic external shell that renders the complex highly soluble in nonpolar lipid regions.

The conformation of the ionophores and their complexes in solution has been studied with several spectroscopic techniques including IR, circular dichroism (CD), and optical rotatory dispersion (ORD), but mainly with ^1H and ^{13}C NMR.

Many studies have attempted to characterize the solution chemistry of ionophore–ion interactions in different solvents in terms of stoichiome-

TABLE I
Representative Antibiotic Ionophores

Group	Name	Structural Type	Ionic Specificity
I Neutral	Valinomycin	Cyclododecadepsipeptide	K^+
	Enniatin	Cyclohexadepsipeptide	K^+
	Antamanide	Cyclodecapeptide	Na^+, Ca^{2+}
	Nonactin	Macrotetrolides	K^+

Valinomycin

Enniatin

Antamanide

Nonactin

II Carboxylic

Nigericin	Poly ether	K^+
Monensin	Poly ether	Na^+
Dianemycin	Poly ether	Na^+
X-206	Poly ether	K^+
Lasalocid (X-537A)	Poly ether	K^+, Ca^{2+}
A-23187	Pyrrole and aminobenzoxazole derivatives of a poly ether-like antibiotic	Ca^{2+}

Nigericin

Monensin

Lasalocid (X-537A)

A-23187

III Channel formers

Gramicidin	Ethanolamides of formyl-pentadecapeptides Icosapeptide	Na^+
Alamethicin	Voltage-dependent pores	

HCO - Val Gly Ala Leu Ala Val Val Val Trp Leu Trp Leu Trp Leu Trp NHCH$_2$CH$_2$OH

1 2 3 4 5 6 7 8 9 10 11 12 13 14 15

L L D L D L L D L D L D L D

Valine - gramicidin A

tries, binding constants, and thermodynamic binding parameters. This has been done using various methods including calorimetry, potentiometry, fluorescence, CD, and NMR spectroscopy. Determination of the kinetics of ion–ionophore binding was, however, only possible using two methods: relaxation studies (Funck *et al.*, 1972) and NMR (Section III,B).

Several important aspects relevant to the ability of ionophores to change the permeability of membranes were investigated in artificial lipid bilayers. Conductance and electric potential measurements in black lipid membranes led to the identification of ionophores either as ion carriers or as channel-forming agents. These studies also provided quantitative data concerning the stoichiometry and charge of the permeating complexes and the relative translocation efficiency of the various ions (for a comprehensive review of this subject see McLaughlin and Eisenberg, 1975 and references cited therein). By employing an ingenious electrical relaxation technique using black lipid membranes it was also possible to determine all the rate constants involved in valinomycin-mediated potassium transport (Gambale *et al.*, 1973; Stark *et al.*, 1971). More recently, NMR measurements on phospholipid multilamellae and vesicles provided preliminary data on ionophore–membrane interactions and the kinetics and mechanism of ionophore-induced permeability (Section IV).

II. Solution Conformation

Nuclear parameters such as chemical shift, the spin–spin coupling constant, T_1 and T_2 relaxation rates, and Overhauser enhancement allow determination of the conformation of molecules in solution. In many cases structural studies based on NMR data can be interpreted with the aid of crystal structure parameters obtained with X-ray analysis. A priori, the crystal structure and structures in solution may differ, and it is therefore important to characterize each structure independently. This has been demonstrated in structural studies on the most extensively documented ionophore, valinomycin, and this molecule therefore will be described in detail in this chapter. It is important, however, to note that NMR conformational studies have been performed for most antibiotic ionophores (Ovchinnikov *et al.*, 1974 and references cited therein). In particular, attention should be drawn to the comprehensive ¹H-NMR studies on polyether ionophores by Anteunis and co-workers (Anteunis, 1981, 1982; Borremans and Anteunis 1981, and references cited therein).

A. ¹H- and ¹³C-NMR Studies

The first reports on the conformation of valinomycin and its alkali ion complexes in solution, deduced primarily from ¹H-NMR data, appeared in 1969 (Haynes *et al.*, 1969; Ivanov *et al.*, 1969; Ohanishi and Urry, 1969). From the ¹H-NMR spectra of free valinomycin and its alkali ion complexes in various solvents (Fig. 1) several immediate conclusions were drawn. (a) Valinomycin and its potassium complex possess a three-fold symmetry axis in solution that markedly simplifies the ¹H spectra. (b) The conformation of free valinomycin is highly dependent on the solvent properties. (c) Complexation to K⁺ is followed by a conformational change to a structure that is almost independent of the solvent.

The first analysis of the spin–spin couplings of the backbone protons (HN—CᵅH) was performed by Ivanov *et al.* (1969). Later the dihedral angles of the backbone and the side chains were calculated from coupling data using modified Karplus relations (Bystrov *et al.*, 1977; Davis and Tosteson, 1975; Ivanov *et al.*, 1971; Patel and Tonelli, 1973). Several heteronuclear spin–spin coupling constants were also measured, $^3J(^{13}C'—N—C^\alpha—^1H)$, $^3J(^1H—NC^\alpha—^{13}C^\beta)$, and $^3J(^1H—C^\alpha C'—^{15}N)$, and then analyzed to give the corresponding dihedral angles from which the torsion angles were derived (Bystrov *et al.*, 1977). In Table II the torsion angles determined directly from the stereochemical dependences of the spin–spin coupling constants are listed in a comparison with X-ray data.

Another important method permitting identification of hydrogen-bonded amides (Kopple *et al.*, 1969) was first used for structural analysis of valinomycin by Ohanishi and Urry (1969). This method is based on measuring the temperature dependence of the chemical shift of amide protons, which in turn is sensitive to hydrogen bonding. Thus a decrease in the slope of the shift indicates hydrogen bonding.

The NMR measurements together with theoretical calculations (Bystrov *et al.*, 1977; Ivanov *et al.*, 1971; Mayers and Urry, 1972; Ohanishi and Urry, 1970; Patel and Tonelli, 1973) allowed characterization of the following main conformations of free valinomycin: (a) form A (according to Patel and Tonelli, 1973) Which is predominant in nonpolar solvents such as heptane and CCl₄ having a "bracelet" conformation in which all six NH groups participate in 4 → 1β-turn hydrogen bonds with the amide carbonyls (Fig. 2a); (b) form B (II) which is predominant in solvents of medium polarity having a "propeller" conformation (Fig. 2b) with only three intramolecular 4 → 1 hydrogen bonds; (c) form C (III) which is characteristic of polar solvents such as aqueous methanol in which valinomycin has no fixed structure and exists as a labile mixture of numerous

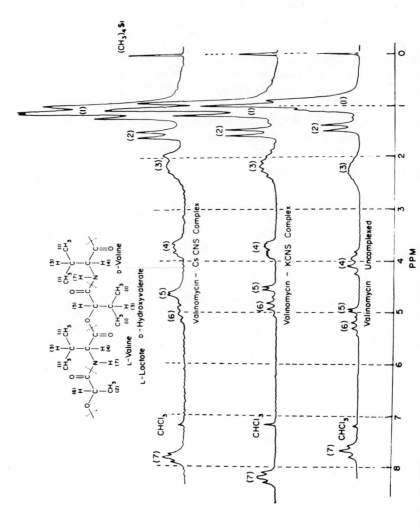

Fig. 1 The 60 MHz ¹H NMR spectra of valinomycin and its KCNS and CsCNS complexes. Valinomycin concentration is 25 mg per ml CDCl₃. The structure of one-third of the molecule is shown in the upper left of the figure, and the numbers on the groups correspond to the numbers of the resonances Haynes et al., (1969).

TABLE II
Torsion Angles of the Preferable Valinomycin Conformation in Various States[a]

Conformational Form	Solvent	Torsion Angles of Residues (degrees)											
		D-Valine			L-Lac		L-Val			D-Hyi		
		ϕ	ψ	χ^1	ϕ	ψ	ϕ	ψ	χ^1	ϕ	ψ	χ^1
K+ complex	C²HCl₃, KCNS crystal	68	-120^b	180	-94	-5	-68	120^b	180	94	5	60
		58.6	-132.7	174.8	-76.2	-12.3	-58.3	130.9	175.7	79.4	3.3	57.5
		57.2	-128.7	176.5	-66.0	-25.1	-60.0	132.7	180.1	86.4	-5.1	61.8
		57.6	-131.2	175.9	-73.1	-15.9	-57.4	132.9	179.0	79.4	8.0	66.3
A	Nonpolar solvents	90	-90^b	180	-120	0	-80	90^b	180	120	0	60
B	CCl₄–(C²H₃)₂SO	110	90^b	-60	-120	120^b	-85	100^b	180	120	0^b	60

[a] (Bystrov et al., 1977)
[b] Evaluated from the molecular model.

257

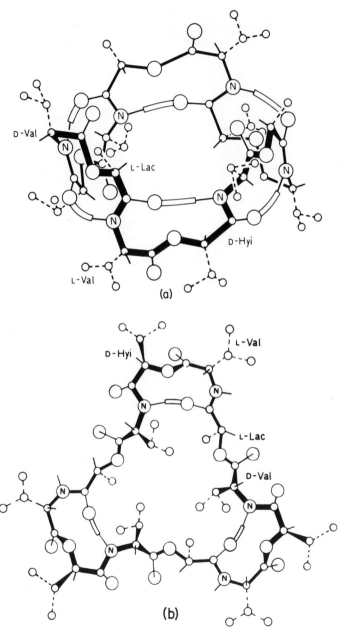

Fig. 2 (a) The predominant "bracelet" conformation of valinomycin in nonpolar solvents. (b) The "propeller" conformation of valinomycin predominant in solvents of medium polarity. Solvent was $CCl_4-(C^2H_3)_2SO$ (3 : 1) (Bystrov *et al.*, (1977).

interconverting conformers. It is important to note that the crystalline structure of free valinomycin as determined by X-ray analysis (Duax *et al.*, 1972; Smith *et al.*, 1975) is different from that of all the conformers found in solution. In the crystalline state, the threefold symmetry is lost and only a pseudosymmetric center is retained. Four out of the six intramolecular hydrogen bonds are regular ($4 \rightarrow 1\beta$ turn), and two are unique ($5 \rightarrow 1$) in closing 13-membered rings. The conformation of the K^+ complex in solution was found to be independent of the solvent, having the form-A bracelet system of hydrogen bonds with additional interaction of the ester carbonyls with the cation to form a cavity 2.8 Å in diameter (Ivanov *et al.*, 1969; 1971; Mayers and Urry, 1972; Ohanishi and Urry, 1970). Unlike free valinomycin, this structure is almost identical to the crystalline structure established by X-ray analysis (Neupert-Laves and Dobler, 1975; Pinkerton *et al.*, 1969) indicating the marked stability of the K^+-valinomycin conformation. ^{1}H- and ^{13}C-NMR studies revealed that the conformation of complexes of valinomycin with the cations Rb^+, Cs^+, and Tl^+ and with ammonium ions closely resembled that of the K^+ complex. On the other hand a different conformation was observed for the Li^+ and Na^+ complexes, which was dependent on the anion and on the polarity of the solvent (Davis and Tosteson, 1975).

The intramolecular ^{1}H nuclear Overhauser effect (NOE) was employed in examining the structure of valinomycin in dimethyl sulfoxide solution (Glickson *et al.*, 1976). The frequency dependence of the NOE indicated positive NOEs at 90 MHz and negative NOEs at 250 MHz. This sign reversal indicates that dipole–dipole interactions are the predominant mechanism of intramolecular NOEs and thus contain geometrical information. An approximate analysis of the NOE data assuming a single correlation time for the entire molecule and ignoring cross-relaxation effects, together with an analysis of the backbone coupling constant data, has shown that the C_1 conformation (the III-1 model of Patel and Tonelli, 1973) is the preferred orientation of valinomycin in dimethyl sulfoxide.

Following the development of ^{13}C NMR, studies on chemical shifts of valinomycin at a natural abundance of ^{13}C provided information complementing the ^{1}H conformational data (Bystrov *et al.*, 1972, 1977; Davis and Tosteson, 1975; Grell and Funck, 1973; Ohanishi *et al.*, 1972; Patel, 1973). The considerable downfield shifts (3–6 ppm) of the ^{13}C signals of the ester carbonyls, due to complexation, served to identify unambiguously the metal-binding sites and to differentiate between the structure of the sodium complex and of the other alkali ion complexes. As mentioned previously, torsion angles were calculated with the aid of additional ^{13}C—^{1}H coupling constants (Bystov *et al.*, 1977).

The use of solvent-induced changes in the ^{1}H and ^{13}C chemical shifts for the various conformations of valinomycin (Fig. 3) has proved to be of particular importance in deducing the conformation of valinomycin in membranes (Section IV).

A specific use of proton magnetic resonance was employed in studying the valinomycin-NH_4^+ complex (Davis, 1975). The T_2 relaxation rate of the ammonium protons determined from line width measurements allowed calculation of the rotational correlation time and the hydrodynamic radius for this particular complex.

B. Metal Nuclide NMR

The chemical shift and relaxation rates of an ion within a complex, and its coupling constants with ligating nuclei, contain structural information, particularly concerning the location of the coordination sites. For all ions there is at least one isotope that possesses a nuclear spin and therefore is amenable to NMR measurements. For example, even the rare isotopes ^{41}K and ^{43}Ca can in practice be monitored by NMR with the recently developed high-field multinuclei pulse–Fourier Transform spectrometers (further examples can be found in the comprehensive review of Forsén and Lindman (1981). However, this approach has so far not been fully exploited, particularly in the area of naturally occurring ionophores. An attempt was made by Haynes et al. (1971) to obtain structural informa-

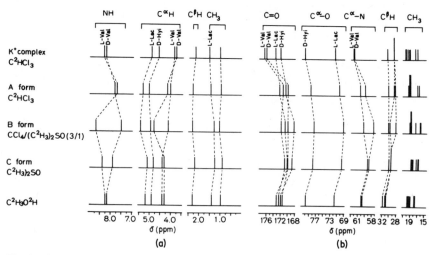

Fig. 3 Changes in (a) ^1H and (b) ^{13}C chemical shifts in the conformational transitions of valinomycin. Values were measured relative to tetramethylsilane (Bystrov et al., (1977).

tion from ^{23}Na chemical shifts and line width measurements for a series of sodium–ionophore complexes. The shifts were found to vary within a range of 25 ppm, but no systematic explanation of their origin was given except for a correlation with the corresponding binding constants, which is probably accidental. The line widths were analyzed for extreme narrowing conditions prevailing for the carrier ionophores. Thus the T_2 relaxation rates of ^{23}Na (derived from the line width) are determined by the correlation time τ_c and by the quadrupolar coupling constant e^2qQ/h. The latter constant originates from the interaction between the quadrupolar moment eQ and the electric field gradients induced by the orientation pattern of the ligands eq. (The ^{23}Na quadrupolar relaxation mechanism is well understood and has been clearly described in several detailed reviews (Civan and Shporer, 1978; Forsén and Lindman, 1981; Laszlo, 1978). Thus by estimating the rotational correlation times from the Debye equation and using crystallographic data and viscosity measurements, the quadrupolar coupling constants could be calculated. The couplings were found to range from 0.5 MHz for monactin to 1.6 MHz for nigericin, in accordance with the symmetry of the ligands around the central ions: four tetrahydrofuran and four carbonyl oxygens, forming an approximately cubic eight-coordination sphere in monactin, and five-coordinated ether, hydroxyl, and carboxyl oxygens with no element of symmetry in nigericin.

The application of ^{205}Tl NMR as a complementary tool for characterizing ionophore complexes was recently successfully demonstrated by Briggs and Hinton (1978a, 1978b, 1979) and Briggs et al. (1980). The thallium ion (Tl$^+$) has an ionic radius and chemical properties similar to those of the potassium ion and can thus substitute for it in complexes and enzymes. The high sensitivity, large chemical shift range, and spectral simplicity of the NMR signal of ^{205}Tl offer clear advantages that encourage its use. The chemical shifts of the Tl$^+$ complexes span a large range (900 ppm) and are summarized together with descriptions of the binding sites in Table III. A qualitative agreement has been found between the chemical shift and the ligand basicity; i.e., the more basic the ligand (such as carboxyl COO$^-$), the larger the low-field shift. The longitudinal relaxation rate was found to be governed by several mechanisms: spin–rotation (SR), dipolar (D), and chemical shift anisotropy (CSA). The dominant mechanism varied, depending on the symmetry of the electronic environment around the metal ion. Thus in the highly symmetric nonactins CSA contributed only 10% (Briggs and Hinton, 1979), whereas in the valinomycin complex relaxation was dominated by CSA, indicating a less symmetric structure (Briggs and Hinton, 1978a). The coupling constants with the ionophore nuclei identify the binding sites and reflect the strength of the

TABLE III

Relationship between the Number and Type of Ligand Atoms and the ^{205}Tl Chemical Shift of Some Tl$^+$-Ionophore Complexes in CHCl$_3$[a]

Ionophore	$\delta(^{205}\text{Tl})$ (ppm)[b]	$\underset{}{>}$C=O	C—O—C	C—OH	$-\text{C}\overset{\displaystyle\text{O}}{\underset{\displaystyle\text{O}^-}{\diagup\hspace{-0.4em}\diagdown}}$
Lasalocid^{-c}	294.5	1 (or 2)	1	2	1 (or 2)
Monensin acid^d	134.2	—	4	2	—
Nigericin^{-e}	127.5	—	4 (or 5)	—	1 (or 0)
Monensin^{-f}	106.6	—	4	2	—
18-Crown-6	−162.0	—	6	—	—
Nonactin	−261.7	4	4	—	—
Monactin	−261.9	4	4	—	—
Dinactin	−262.2	4	4	—	—
Valinomycin	−540.5	6	—	—	—

Number and type of ligands bound per hour

[a] (Briggs *et al.*, 1980)
[b] For 0.1 *M* solutions at 24°C. $\delta = 0$, ^{205}Tl$^-$ in H$_2$O at infinite dilution.
[c] A second carbonyl oxygen probably binds at high concentrations ($\delta = 288$ ppm) and a second carboxylate oxygen at low concentrations ($\delta = 298$ ppm).
[d] One of the hydroxyl groups is probably hydrogen-bonded to the ClO$_4^-$ anion, rendering that hydroxyl oxygen more basic than in the monensin$^-$ complex.
[e] At high concentrations ($\delta = 100$ ppm), binding is probably by five ethers and possibly one carboxylate, weakly; at low concentrations ($\delta = 140$ ppm), binding is probably by four ethers and one carboxylate, strongly, and possibly another ether, weakly.
[f] Another signal is also present, at 96.7 ppm.

electronic metal–ligand overlap. ^{13}C satellites due to coupling of Tl$^+$ with the carbonyl ionophore nuclei of valinomycin were observed. These satellites confirmed the occurrence of strong metal–ligand interactions, also indicated by ^{13}C-NMR studies (Bystrov *et al.*, 1977).

III. Nuclear Magnetic Resonance Studies on Ion–Ionophore Interactions in Solution

Carrier transport involves the formation and dissociation of an ion–ionophore complex, either in solution or at the interface of the membrane and the solution. Similar interactions occur between cations and the inner surface of channel-forming antibiotics, although these ionophores require an additional orienting contribution from the membrane. In most cases the solvents used in these studies were selected to mimic the properties of both the polar interface and the hydrophobic interior of membranes.

In order to present the data in a uniform mode we shall use the following notation: I, a neutral ionophore; IH, the acid form of a carboxylic ionophore; I^-, the ionized form of a carboxylic ionophore; M^{n+}, the ion and its charge.

Concentrations are indicated by square brackets [ ]. The subscript t designates total concentration, P, partial, and F and B, free and bound species respectively. We use conventional notation for the various NMR and common physical and chemical parameters.

A. Equilibrium and Thermodynamic Parameters

When NMR signals due to free and complexed nuclei appear at separate frequencies, equilibrium binding parameters can be readily determined from their corresponding integrated areas. This trivial approach can not be employed for ionophore–ion complexes, because in most cases the lifetimes of these complexes τ_B are very short and only averaged signals can be observed. At the fast-exchange limit, the NMR chemical shifts and relaxation rates are a weighted average of these properties in the free and complexed species, from which the information needed for determining binding constants can be obtained. For a fast reaction of the type $I + M^+ \rightleftharpoons MI^+$, with an equilibrium concentration constant $K = [MI^+]/[I][M^+]$, the complexation-induced shift of either the ionophore nuclei or the ion nucleus is given by the following equation derived in this form by Prestegard and Chan (1969):

$$\delta_B - \delta_F = \tfrac{1}{2}(\delta_B - \delta_F)(1 + \phi + \eta) - [(1 + \phi + \eta)^2 - 4\phi]^{1/2} \quad (1)$$

When the ionophore nuclei are monitored,

$$\phi = [M^+]_t/[I]_t \quad \text{and} \quad \eta = 1/K[I]_t$$

And when the ion nucleus is monitored,

$$\phi = [I]_t/[M^+]_t \quad \text{and} \quad \eta = 1/K[M^+]_t$$

concentration-dependent measurements of the shift can then be analyzed with the aid of a nonlinear best-fit procedure to yield the binding constant K and the chemical shift in the bound state δ_B.

Prestegard and Chan (1969; 1970) employed this method to study the binding properties of the ionophore nonactin with the alkali ions sodium, potassium, and cesium in both dry and wet acetone. In dry acetone all three ions bind to nonactin with nearly equal affinity, however, the addition of water markedly reduced only the binding constants for Na^+ and Cs^+, making the binding of K^+ highly favored. It was suggested that subtle differences in the variations of the standard free energies of ion hydration and ion complexation with ion size were responsible for this selectivity (Prestegard and Chan, 1970).

An approach similar in principle to that described previously, but based on measuring weighted average relaxation rates of the ion nucleus, was employed by Cornelis and Laszlo (1979) in a study on sodium binding to gramicidin A. The ^{23}Na line width was measured at two frequencies (23.8 and 62.86 MHz) as a function of the gramicidin/sodium molar ratio. The gramicidin (G) was assumed to be present as a dimer (G_2) to which sodium ions bind according to $G_2 + Na^+ \rightleftharpoons G_2,Na^+$, with a concentration equilibrium constant $K = [G_2,Na^+]/[G_2][Na^+]$. The data were analyzed according to Eq. (1) (for the ion nucleus), except that chemical shifts were replaced by the corresponding line widths. Such a replacement is valid only if the difference in chemical shift between the bound and free states is negligible compared to the corresponding line width difference. On the other hand, this replacement is always correct when using T_1 relaxation rates. It should be noted that relaxation rates (hence line widths) are affected by the viscosity, which in turn depends on the concentration of the solute. The apparent relaxation rates should therefore be normalized to a standard viscosity.

^{23}Na line width measurements were also used to determine the binding constant of sodium ions to X-537A (Grandjean and Laszlo, 1979). In this study the technique was extended to competitive binding studies on several biogenic amines. The reduction in the ^{23}Na line width due to the displacement of sodium cations from the complex by amine cations was followed as a function of the amine concentration.

Some ionophores can form complexes with variable stoichiometries. For instance the stoichiometry of the complexes of X-537A depends on the ion and on the solvent (Degani and Friedman, 1974). In principle the various species can be identified by determining the different binding constants. An example of this is the study of the complexes of Pr^{3+} with X-537A in methanol, using CD, fluorescence, and NMR shift measurements of the ionophore protons (Chen and Springer, 1978). The results indicated that 1:1, 1:2, and 1:3 Pr^{3+}:X-537A species can exist in solution. The corresponding three binding constants were calculated with an accuracy of one order of magnitude using a nonlinear least squares program.

Table IV summarizes all the NMR equilibrium binding studies.

TABLE IV

Ionophore–Ion Binding Constants as Determined by NMR Measurements

Ionophore	Ion	Solvent	K at $25°C^a$ (M^{-1})	Method	Reference
Valinomycin	Na^+	Acetone	25	^1H shift	Davis and Tosteson (1975)
	K^+	Acetone	$>10^4$	^1H shift	
Nonactin	Na^+	Acetone	7.10^4	^1H shift	Prestegard and Chan (1970)
	K^+	Acetone	7.10^4	^1H shift	Prestegard and Chan (1970)
	Cs^+	Acetone	$1.5.10^4$	^1H shift	Prestegard and Chan (1970)
	Na^+	Wet acetone	$2.1.10^2$	^1H shift	Prestegard and Chan (1970)
	K^+	Wet acetone	$1.7.10^4$	^1H shift	Prestegard and Chan (1970)
	Cs^+	Wet acetone	4.10^2	^1H shift	Prestegard and Chan (1970)
Lasalocid	Pr^{3+}	Methanol	10^7	^1H shift	Chen and Springer (1978)
	Pr^{3+}	Methanol	10^{6b}	^1H shift	Chen and Springer (1978)
	Pr^{3+}	Methanol	10^{5c}	^1H shift	Chen and Springer (1978)
	Na^+	Methanol–hexane, 71:29	500	^{23}Na T_2 relaxation	Grandjean and Laszlo (1979)
	Serotonin bimaleate	Methanol–hexane, 71:29	450	^{23}Na T_2 relaxation by competition	Grandjean and Laszlo (1979)
	3-Hydroxytyramine	Methanol–hexane, 71:29	260	^{23}Na T_2 relaxation by competition	Grandjean and Laszlo (1979)
	L-Norepinephrine	Methanol–hexane, 71:29	280	^{23}Na T_2 relaxation by competition	Grandjean and Laszlo (1979)
Gramicidin A	Na^+	Ethanol–water, 90:10	4.0^d	^{23}Na T_2 relaxation	Cornelis and Laszlo (1979)

[a] For reactions of the type $M + I \rightleftharpoons MI$.
[b] For a reaction of the type $MI + I \rightleftharpoons MI_2$.
[c] For a reaction of the type $MI_2 + I \rightleftharpoons MI_3$.
[d] For a reaction of the type $M + I_2 \rightleftharpoons MI_2$.

B. Kinetics

The rates of ionophore–ion association and dissociation in solution are usually very fast and therefore readily accessible to magnetic resonance spectroscopy. The basis for analyzing the kinetics of a process that involves the exchange of nuclei among several environments that differ in magnetic properties has been extensively reviewed (Buckley *et al.*, 1975; Kaplan and Fraenkel, 1980; Loewenstein and Connor, 1962). Here we shall describe the specific approaches used to elucidate the kinetics of the complexation of ionophores with ions in solution.

In systems containing either free and ion-bound ionophore or free and ionophore-bound ions two different exchange mechanisms can occur (Haynes, 1972). The first is a dissociative mechanism of the general type

$$I + M \underset{k_{off}}{\overset{k_{on}}{\rightleftharpoons}} MI$$

The second is a transfer mechanism of either the ionophore

$$I^* + MI \overset{k_2}{\rightleftharpoons} I + MI^*$$

or the ion

$$M^* + MI \overset{k_2}{\rightleftharpoons} M^*I + M$$

A NMR kinetic analysis enables one to determine the residence time of a nucleus in either the free τ_F or bound τ_B state, which at equilibrium are related to one another by the equation

$$\tau_F^{-1}[I] = \tau_B^{-1}[MI] \tag{2}$$

above. These residence times can be expressed in terms of the above rate constants. For the bound ionophore nuclei:

$$\tau_B^{-1} = k_{off} + k_2[I] \tag{3a}$$

and for the bound ion nucleus:

$$\tau_B^{-1} = k_{off} + k_2[M] \tag{3b}$$

Thus concentration-dependent studies on either lifetime can provide information on the mechanism.

Haynes (1972) and Haynes *et al.* (1969) were the first to determine kinetic data from complexation-induced line broadenings of protons in valinomycin and nonactins. Measurements were obtained with high ratios of free to complexed ionophore molecules. Under these conditions the following basic equations for the shift δ and line width $\Delta H_{1/2}$ apply (Johnson, 1965).

$$\delta = \delta_F + \frac{(\delta_B - \delta_F)[MI]}{1 + (\delta_B - \delta_F)^2 \tau^2 [I]} \tag{4a}$$

$$\Delta H_{1/2} = (\Delta H_{1/2})_F + \frac{[MI]}{[I]_t} \frac{(|\delta_B - \delta_F|)}{1 + (|\delta_B - \delta_F|)^2 \tau^2} \tag{4b}$$

where τ^{-1} is approximately equal to τ_B^{-1}. It should be noted that Eq. (4b) covers the entire range from very fast ($\delta_B - \delta_F \ll 1/\tau_B$) to very slow ($\delta_B - \delta_F \gg 1/\tau_B$) exchange rates. This equation was further simplified by Haynes (1972) to a relation that holds for slow-exchange behavior. Concentration-dependent line broadening measurements revealed that the exchange was dominated by the dissociative mechanism from which approximate values for the dissociation rate constants (Table V) could be obtained.

An alternative method based on measurements of nuclear relaxation rates of the complexed cation (e.g., ^{23}Na) was applied successfully in studying the kinetics of sodium binding to valinomycin (Shporer et al., 1974) and to monensin (Degani, 1977) in methanol (Fig. 4). The same method had been employed earlier by Schori et al. (1971) to measure sodium exchange with crown ethers. It is based on the fact that the nuclear relaxation rates ($1/T_1$ and $1/T_2$) of sodium bound to the ionophore are much faster than that of free sodium. (Here and elsewhere "free" means the solvated uncomplexed ion.) In solutions containing both complexed and free ions, transfer between the two states at a rate comparable to the difference between the relaxation rates affects both relaxations. Usually it is easier to measure the enhancement in the rates of relaxation of free sodium. An example of the exchange of free sodium with a monensin-Na$^+$ complex is shown in Fig. 4. The lifetime is calculated according to the equation derived by Schori et al. (1971) following Woessner's theoretical analysis (Woessner, 1961):

$$\frac{1}{\tau_F} = \frac{(1/T_{1B} - 1/T_1)(1/T_1 - 1/T_{1F})P_B}{1/T_{1av} - 1/T_1} \tag{5}$$

where $1/T_{1av} = P_F T_{1F}^{-1} + P_B T_{1B}^{-1}$. The same equation holds for T_2 relaxation rates on the condition that the difference between the chemical shifts of the complexed and free ions is smaller than the difference between their rates of relaxation.

Concentration dependence studies on τ_F^{-1} revealed that for valinomycin as well as for monensin the dominant mechanism was via the first-order dissociative exchange I + M $\rightleftharpoons$ MI. The activation parameters for this exchange were obtained from measurements of the dependence of the rate on temperature. By comparing the kinetic data obtained for several ionophores it was noted (Degani, 1977) that the binding rate constants for several ionophore–ion complexes differed by no more than an order of magnitude (in the same solvent) and were close to diffusion-controlled rates. On the other hand, the dissociation rate constants varied by several

TABLE V

Kinetic Data for the Ion–Ionophore Interaction in Solution

Ionophore	IOH	Solvent	$1/\tau_B$ at 25°C $(s^{-1})^a$	$1/\tau_F$ at 25°C $(s^{-1} M^{-1})$
Nonactin	K^+	Methanol–chloroform, 80:20	32	—
Monactin	K^+	Methanol–chloroform, 80:20	23	—
Dinactin	K^+	Methanol–chloroform, 80:20	21	—
Trinactin	K^+	Methanol–chloroform, 80:20	18	—
Valinomycin	K^+	Methanol–chloroform, 80:20	21	—
	K^+	Acetone	8.5	—
	K^+	Chloroform	—	—
	Na^+	Chloroform	—	—
	Na^+	Methanol	2.10^6	—
Monensin	Na^+	Methanol	63	6.3×10^7
Lasalocid	Ni^{2+}	Methanol	2.4×10^2	2.2×10^6
	Mn^{2+}	Methanol	$>1.4 \times 10^4$	$>3.5 \times 10^8$
	Co^{2+}	Methanol	1.2×10^4	4×10^8
	Pr^{3+}	Methanol	2.5×10^{5b}	—
	Ba^{2+}	Methanol	5.2×10^3	1.5×10^{10}
	Ba^{2+}	Methanol	7.6×10^3	—
	Sr^{2+}	Methanol	2.9×10^4	—
	Ca^{2+}	Methanol	4.1×10^4	—
	K^+	Methanol	6.6×10^4	—
	Ba^{2+}	Chloroform	2×10^{3c}	—
	2-Aminoheptane	Chloroform	80	—
	1-Amino-2-phenylethane	Chloroform	80	—
	Dopamine	Chloroform	4×10^2	—
	Norepinephrine	Chloroform	4.3×10^2	—
Gramicidin A	Na^+	Ethanol–water, 90:10	$<5.5 \times 10^{8d}$	$<2.2 \times 10^9$

TABLE V

Kinetic Data for the Ion–Ionophore Interaction in Solution

$\Delta H^\dagger$ (kcal/mol)	$\Delta S^\dagger$ (eu)	$\Delta E^\dagger$ (kcal/mol)	Method	Ref.
—	—	—	^{1}H exchange broadening	Haynes (1972)
—	—	—	^{1}H exchange broadening	Haynes (1972)
—	—	—	^{1}H exchange broadening	Haynes (1972)
—	—	—	^{1}H exchange broadening	Haynes (1972)
—	—	—	^{1}H exchange broadening	Haynes (1972)
—	—	29	^{1}H exchange broadening	Davis and Tosteson (1975)
—	—	13	^{1}H exchange broadening	Davis and Tosteson (1975)
—	—	14, 10	^{1}H exchange broadening	Davis and Tosteson (1975)
—	—	9.5	^{23}Na T_1 relaxation	Shporer et al. (1974)
10.3	−15.8	—	^{23}Na T_1 relaxation	Degani (1976)
4.5	−32	—	^{1}H paramagnetic exchange broadening	Degani and Friedman (1975)
—	—	—	^{1}H paramagnetic exchange broadening	Degani and Friedman (1975)
1.9	−33	—	^{1}H paramagnetic exchange broadening	Degani et al. (1976)
—	—	—	^{1}H paramagnetic exchange broadening	Chen and Springer (1978)
6.5	−20	—	^{1}H paramagnetic exchange broadening	Shen and Patel (1976)
—	—	—	^{1}H (see text)	Krishnan et al. (1978)
—	—	—	^{1}H (see text)	Krishnan et al. (1978)
—	—	—	^{1}H (see text)	Krishnan et al. (1978)
—	—	—	^{1}H (see text)	Krishnan et al. (1978)
—	—	10.2	^{1}H exchange broadening	Patel and Shen (1976)
—	—	—	^{1}H exchange broadening	Shen and Patel (1977)
—	—	—	^{1}H exchange broadening	Shen and Patel (1977)
—	—	—	^{1}H exchange broadening	Shen and Patel (1977)
6.8	−23.8	—	^{1}H exchange broadening	Shen and Patel (1977)
—	—	—	^{23}Na linewidth studies	Cornelis and Laszlo (1979)

[a] For reactions of the type $M + I \rightleftharpoons MI$.
[b] For a reaction of the type $MI_2 + I \rightleftharpoons MI_3$.
[c] For a reaction of the type $MI^* + HI \rightleftharpoons MI + HI^*$.
[d] For a reaction of the type $M + I_2 \rightleftharpoons MI_2$.

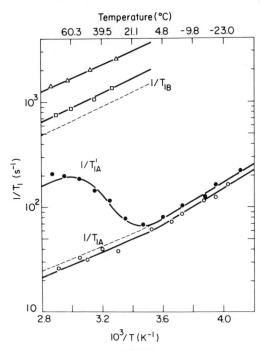

Fig. 4 Temperature dependence of the longitudinal relaxation rates of Na^+ in methanol in the presence of monensin. Solutions composition: ○ 0.5 M NaBr, ● 0.15 M MonNa and 0.35 M Na NaBr, □ 0.3 M MonNa, △ 0.65 M MonNa; $1/T_{1B}$(---) 0.15 M MonNa extrapolated from higher concentrations; $1/T_{1A}$(---) 0.5 M NaBr corrected for the change in viscosity; $1/T'_{1A}$(——) calculated (Degani, 1977).

orders of magnitude. Thus, in the latter case, specificity is apparently mainly due to differences in dissociation rates.

A novel approach, suitable for measuring very fast kinetics, has been used by Cornelis and Laszlo (1979) to study the binding of sodium ions to gramicidin A dimers. This method is, in general, suitable for nuclei with a spin larger than 1 that are attached to a slowly reorienting macromolecule part of the time. For instance, the transverse relaxation rate of sodium ions $1/T_2$ spending a fraction of time τ_B bound to a large molecule with a slow correlation time for rotation τ_c is given by the following equations (Delville *et al.*, 1979).

$$1/T_2 = 1 - P_B/T_{2F} + P_B[0.6/(T'_{2B} + \tau_B) + 0.4/(T''_{2B} + \tau_B)] \quad (6a)$$

$$1/T'_{2B} = \pi/5(e^2qQ/h)^2[\tau_c + \tau_c/(1 + \omega^2\tau^2)] \quad (6b)$$

$$1/T''_{2B} = \pi^2/5(e^2qQ/h)^2[\tau_c/(1 + 4\omega^2\tau_c^2) + \tau_c/(1 + \omega^2\tau_c^2)] \quad (6c)$$

In the experiment on sodium exchange with gramidicin dimer the system obeys the following conditions:

$$\tau_B \ll T'_{2B}, T''_{2B} \qquad P_B \ll 1$$

Thus Eq. (6a) can be approximated to

$$1/T_2 = 1/T_{2F} + P_B(0.6/T'_{2B} + 0.4/T''_{2B}) \tag{7}$$

Because T'_{2B} and T''_{2B} are frequency-dependent [Eqs. (6b) and (6c)], relaxation measurements at two frequencies (23.81 and 62.86 MHz) allowed τ_c to be determined. It has been further assumed that τ_c is predominately determined by the residence time τ_B, from which a decomplexation constant of about $(5.5 \pm 0.5) \times 10^8$ s^{-1} has been derived. This value, together with the independent determination of the equilibrium binding constant, implies an association rate constant of $(2.2 \pm 0.2) \times 10^9$ that is very fast and close to the diffusion-controlled limit. These data reinforce the suggestion that specific cation-binding sites are located at each entrance to the gramicidin A channel, in addition to binding sites inside the channel.

^{1}H-NMR measurements of an ionophore can be applicable to studies on the kinetics of ion complexation if the chemical shift and/or broadening of the ^{1}H signals in the complexed state is sufficiently large. Such a situation occurs when complexation to paramagnetic ions takes place. The kinetics of X-537A interaction with the ions Ni^{2+}, Mn^{2+}, and Co^{2+} in methanol solution (Degani and Friedman, 1975; Degani et al., 1976) was studied by following three of the X-537A protons (H-4, H-5, H-29, Table I) that are markedly affected by binding. The data were analyzed using the Swift–Connick equations for exchange (Swift and Connick, 1962), expressed by Degani and Friedman (1975) in a rate form ($k_2 = 1/T_2$; $k_B = 1/\tau_B$)

$$k_2 = k_B \frac{k_{2B}(k_{2B} + k_b) + \Delta\omega_B}{k_{2B} + k_B + \Delta\omega_B{}^2} \tag{8}$$

$$s = \frac{k_B\Delta\omega_B}{(k_{2B} + k_B)^2 + \Delta\omega_B{}^2} \tag{9}$$

where $k_2 = (1/T_2 - 1/T_{2F})[IH]/[MI]$ and $s = (\delta - \delta_F)[IH]/[MI]$.

These equations were further simplified for the limiting cases of fast exchange, $k_B \gg k_{2B}, \Delta\omega_B$, and slow exchange, $k_{2B} \gg \Delta\omega_B, k_B$ or $\Delta\omega_B \gg k_{2B}, k_B$. Rate constants and activation parameters for Ni^{2+} and Co^{2+} complexes and a limit for the rate constant of the Mn^{2+} complex were obtained (Table V). The main difference between these paramagnetic complexes and the other alkali and alkaline earth complexes is that their equilibrium binding constants are very similar but their selectivity is expressed by large differences in the rates of association and dissociation.

A similar method was applied in measuring the lifetime of the ternary X-537A complex of praseodymium ion (PrX_3) in methanol (Chen and Springer, 1978). The line broadening data for X-537A protons were analyzed according to the formulation of Lenkinski and Reuben (1976) applicable to fast-exchange processes involving the lanthanide ions:

$$(1/T_2 - 1/T_{2F})P_B = 1/T_{2B} - 1/T_{2F} + (1 - P_B)^2 \tau_B \Delta\omega_B^2 \qquad (10)$$

τ_B can be calculated from the slope of a plot of the normalized line broadening $(1/T_2 - 1/T_{2F})/P_B$ versus $(1 - P_B)^2$ and by independent measurements of $\Delta\omega_B$ (Section III,A). For this complex $\tau_B = 3.9 \mu s$, which is rather short relative to the lifetimes of complexes with other ions (Table V) but much longer than the water residence time in the lanthanide hydration sphere ($5 \times 10^{-3} \mu sec$ at 25°C).

The fast-exchange conditions in the X-537A–Pr^{3+} system and the large shifts of X-537A protons (H-4, H-5, and H-29 primarily) induced in the presence of this ion, led to an elegant use of this system (Krishnan *et al.*, 1978) as a means of measuring the kinetics of complexation with diamagnetic ions that cannot be measured directly (although in certain cases using high magnetic fields, may permit direct measurement, as will be seen later in this section). In this study the lifetime of PrX_3 in methanol was determined again by measurements in solutions containing a small fraction of the complex and using Swift–Connick equations. The main aim of this study, however, was to determine the lifetime of the X-537A complexes with the alkaline and alkaline earth ions using the effects of PrX_3 on the shifts and relaxations of the ionophore protons. Equations for a three-site exchange system were derived: site 1, the ionophore bound to the Pr^{3+}; site 2, free ionophore; site 3, the ionophore bound to the diamagnetic metal. Measurements of solutions containing fixed fractions of free and metal-complexed ionophore and varied amounts of Pr^{3+} complex yielded lifetimes for the complexes of Ba^{2+}, Sr^{2+}, Ca^{2+}, and K^+ (Table V). The lifetimes for Na^+ and Li^+ complexes were too short to be accurately determined.

In polar solvents the proton shifts due to complexation with a diamagnetic ion are usually not large enough relative to the rate of exchange to allow kinetic analysis. However, by working at a high magnetic field (84 kG), Shen and Patel (1976) were able to shift the H-10 of the Ba^{2+} complex of X-537A in methanol to a range suitable for kinetic studies. A computer program written by Kleier and Binsch (1970) was used to generate the line shapes as a function of the exchange rate for given values of chemical shifts, coupling constants, and relaxation times. Temperature-dependent measurements yielded the activation parameters, and concentration-dependent studies indicated that the exchange proceeds via a unimolecular dissociation mechanism. In less polar solvents the exchange rates are slowed down sufficiently to fall within a range accessible for ^{1}H-NMR kinetic studies. Davis and Tosteson (1975) measured the kinetics of potassium and sodium complexation to valinomycin in chloroform and in acetone by employing complete exchange-modified line shape analysis. (See reviews of Buckley *et al.*, 1975; Loewenstein and Connor, 1962). Patel and Shen (1976) have measured the rates and activation energy for the exchange of the Ba^{2+} complex of X-537A (a 1:2 ion–ionophore complex) in chloroform with the free acidic form of this ionophore. The line shape of H-4 and H-5 of X-537A were studied as a function of temperature and were analyzed by simulation (Kleier and Binsch, 1970), yielding the corresponding rate constants shown in Fig. 5. Concentration-dependent measurements showed that, unlike the situation in polar solvents, a bimolecular transfer mechanism dominates ion exchange in nonpolar solvents. In a similar study using amine complexes of X-537A in chloroform (Shen and Patel, 1977) both mechanisms were found to contribute to the exchange, although the contribution varied depending on the nature of the amine studied.

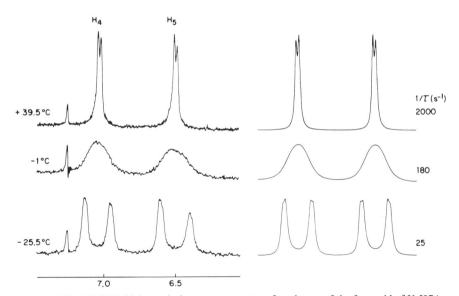

Fig. 5 The 360-MHz high-resolution proton spectra of a mixture of the free acid of X-537A and the barium complex (BaX_2) in [^{2}H] chloroform between 6.5 and 7.5 ppm at $-25.5°C$, $-1°C$ and 39.5°C. The adjoining spectra represent simulations for various values of $1/\tau$, where τ is the lifetime in each state before exchange (Patel and Shen, 1976).

Table V summarizes all the solution kinetic data obtained so far for the ionophores.

IV. Nuclear Magnetic Resonance Studies on Ionophores in Membrane Systems

The activity of ionophores as mediators of ion transport is manifested in membrane systems. It is therefore of prime importance to characterize the structure and mechanism of ionophore operation in such systems.

A. Effects on Membrane Structure

The insertion of a lipophilic molecule such as an ionophore into a lipid bilayer may affect the order and the dynamics of the lipid molecules in the bilayer, particularly in regions close to the ionophore. By monitoring changes in the NMR spectrum of the lipid nuclei, ^1H, ^2H, ^{13}C, and ^{31}P, it should be possible to determine the extent and location of the perturbation.

The first observations of this kind were made by Finer et al. in 1969. The effect of the antibiotics alamethicin, valinomycin, and gramicidin S on sonicated phospholipid vesicles was studied by monitoring membrane protons with NMR. Insertion of valinomycin and alamethicin (at a lipid/ionophore molar ratio of 100:1) caused a large broadening in the signal of the chain hydrocarbon protons and a partial broadening of the methylcholine resonance. This broadening was interpreted as arising from a reduction in the motion of the lipid molecules, leading to increased correlation times and hence to reduced relaxation times. In a later study, the same authors (Hauser et al., 1970) extended their work with alamethicin. At a lipid/alamethicin molar ratio of up to 50:1 the intensity of all the membrane proton signals went down, but their area remained constant. Below this ratio a concomitant decrease in the proton signal area of the membrane was observed. It was concluded that alamethicin interacts primarily with the lipid chains in the vesicles. However, similar studies by Chan and co-workers (Hsu and Chan, 1973; Lau and Chan, 1974, 1975) led to a different conclusion, namely, that the perturbation by ionophores such as valinomycin and alamethicin is confined to the polar head group region at the bilayer–water interface. The latter studies were performed with multilamellar dispersion employing a delayed Fourier transform (DFT), which involves introduction of a receiver dead time between the end of the rf pulse and the start of data collection in order to filter out the broad components and isolate the sharp ones (Chan *et al.*, 1971). The addition of valinomycin to multilamellar lipid dispersion (lipid/ionophore ratio up to 50 : 1) induces a reduction in the intensity and a narrowing in the line width of the proton signal of the methylcholine groups. The drop in intensity could be explained as being due to a slow equilibrium between bound and free sites, the signal of the bound site being broadened beyond detection. The line width narrowing was attributed to enhancement of the segmental motion of the nonbound polar head groups. Hsu and Chan (1973) suggested that the discrepancy between the induced effects on multilamellar dispersions and in sonicated vesicles (Finer *et al.*, 1969) stemmed from differences between the properties of multilamellar membranes and the highly curved vesicular membranes. The interaction of alamethicin with multilamellar membranes is bizarre; an increase in the intensity of the sharp component of the methylcholine signal is observed. Moreover, when added to sonicated vesicular membranes, alamethicin causes fusion of the vesicles (Lau and Chan, 1974, 1975), as reflected by line broadening and a reduction in the intensity of all proton signals. The fusion process was followed by monitoring the inner and outer methylcholine protons, employing eropium ions as a shift reagent. As a result of the fusion the shift between the inner and outer methylcholine signals decreased gradually until a final collapse to one signal occurred.

The complexity of the proton spectra of lipid membranes limited analysis of the data to qualitative conclusions. The employment of deuterium quadrupolar measurements in membrane studies enables one to use a quantitative approach, namely, direct determination of the ordering of the lipids within the membrane. The disordering effect caused by the insertion of gramicidin A into a bilayer was inferred from complete collapse of the ^2H quadrupolar splitting of lipids deuterated at the end methyl groups of the lipid chains (Oldfield *et al.*, 1978). These measurements were performed at a very high ionophore/lipid ratio (67 wt% gramicidin A). Although they may be useful for analyzing lipid–peptide interactions in general, they do not reflect the effects of the ionophores at very low concentrations. A similar system was studied by Cornell *et al.*, (1980), employing ^1H—^{13}C cross-polarization measurements. The presence of gramicidin shortened the cross-polarization time of the lipid methylene protons and their spin–lattice relaxation time T_1 in the rotating frame, suggesting a decrease in the amplitude of the chain's motion and an increase in the relative intensity of the low-frequency motion of the chains, respectively. Evidently the results obtained so far are conflicting and inconclusive, and further studies are needed to define the effects exerted by ionophores on membrane structure.

B. *Organization within the Membrane*

Section II described conformational studies in homogeneous solutions. Extension of these studies to membranes is essential for a final confirmation of the structure–activity relationships. Two approaches were employed to eliminate the masking effect of signals due to the membrane nuclei. In one set of experiments (Feigenson *et al.*, 1977; Feigenson and Meers, 1980) the ¹H-NMR spectra of the ionophores were studied in perdeuterated lipid membranes. In a second set of experiments (Weinstein *et al.*, 1979, 1980) the ionophore itself was specifically labeled with ¹³C and ¹⁹F nuclei.

In a preliminary study, Feigenson *et al.* (1977) monitored the proton tryptophan signals of gramicidin A embedded in lipid bilayers. However, because of the difficulty in peak assignment, no structural information could be obtained.

Recently Feigenson and Meer (1980) recorded and fully assigned the ¹H spectra of valinomycin and its alkali ions complexes in sonicated vesicles. The signals were broadened to 20–40 Hz, thus preventing spin–spin splittings from being determined. On comparing the proton chemical shifts of valinomycin obtained in the membrane and in solution (Fig. 3) these authors suggested that the conformation of both uncomplexed and complexed ionophores are similar to those determined in nonpolar solvents such as cyclohexane and chloroform (see Section II for relevant references). On titrating with alkali ions, one averaged signal was observed for each proton, providing a lower limit for the exchange rate of >150 s^{-1}. The chemical shift data suggest that valinomycin is mainly located within the hydrocarbon region, whereas the kinetic limit seems to suggest a polar location at the interface.

Clear-cut results were obtained in a study on gramicidin A embedded in lipid vesicles (Weinstein *et al.*, 1979; 1980). The isotopically labeled derivatives in Fig. 6 were synthesized and used in this study. All were found to induce conductivity in planar lipid bilayers, indicating a retention of their ionophoretic properties. The labeled nuclei were monitored in the presence of the shift and relaxation reagents in the water phase, thamarium ions (Tm^{3+}), and manganous ions (Mn^{2+}), respectively. The nitroxide spin-labeled derivative of phosphatidylcholine was used in the bilayer phase. It was expected that the paramagnetic ions would interact with nuclei of gramicidin located in the vicinity of the interface, whereas the spin-labeled lipid should exert its maximal effect on nuclei close to the central hydrocarbon core. The results summarized in Fig. 7 clearly show that the enhanced shift and relaxation by the paramagnetic ions are observed for carbons near the carbonyl groups, indicating that the C-termi-

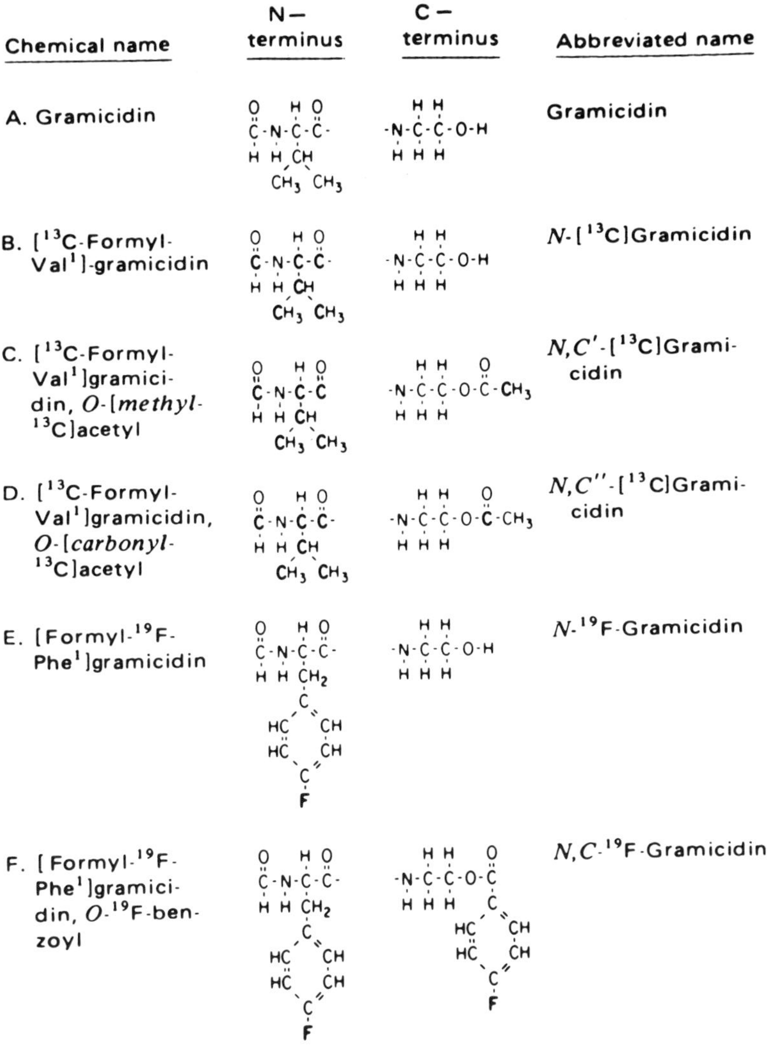

Chemical name	N— terminus	C— terminus	Abbreviated name

A. Gramicidin

N-terminus:
```
O   H O
C-N-C-C-
H   H CH
      CH₃  CH₃
```
C-terminus:
```
 H H
-N-C-C-O-H
 H H H
```
Abbreviated name: Gramicidin

B. [¹³C-Formyl-Val¹]-gramicidin

N-terminus:
```
O   H O
C-N-C-C-
H   H CH
      CH₃  CH₃
```
C-terminus:
```
 H H
-N-C-C-O-H
 H H H
```
Abbreviated name: N-[¹³C]Gramicidin

C. [¹³C-Formyl-Val¹]gramicidin, O-[methyl-¹³C]acetyl

N-terminus:
```
O   H O
C-N-C-C
H   H CH
      CH₃  CH₃
```
C-terminus:
```
 H H   O
-N-C-C-O-C-CH₃
 H H H
```
Abbreviated name: N,C'-[¹³C]Gramicidin

D. [¹³C-Formyl-Val¹]gramicidin, O-[carbonyl-¹³C]acetyl

N-terminus:
```
O   H O
C-N-C-C-
H   H CH
      CH₃  CH₃
```
C-terminus:
```
 H H   O
-N-C-C-O-C-CH₃
 H H H
```
Abbreviated name: N,C''-[¹³C]Gramicidin

E. [Formyl-¹⁹F-Phe¹]gramicidin

N-terminus:
```
O   H O
C-N-C-C-
H   H CH₂
        C
      HC   CH
      HC   CH
        C
        F
```
C-terminus:
```
 H H
-N-C-C-O-H
 H H H
```
Abbreviated name: N-¹⁹F-Gramicidin

F. [Formyl-¹⁹F-Phe¹]gramicidin, O-¹⁹F-benzoyl

N-terminus:
```
O   H O
C-N-C-C-
H   H CH₂
        C
      HC   CH
      HC   CH
        C
        F
```
C-terminus:
```
 H H   O
-N-C-C-O-C
 H H H    C
        HC   CH
        HC   CH
          C
          F
```
Abbreviated name: N,C-¹⁹F-Gramicidin

Fig. 6 Chemical structure and nomenclature of labeled gramicidin analogs and derivatives. The boldface C denotes specific ¹³C enrichment. Only the first amino acid residue is indicated at the NH₂ terminus, and only the ethanolamide and its esters are indicated on the COOH terminus. The intervening 14 amino acids are not shown (Weinstein *et al.*, 1979).

nal of the gramicidin A channel is located near the membrane surface. The enhanced effect of the spin label on the N-terminal valine methyls indicates the formation of an N-terminal–N-terminal helical dimer with the N-terminal buried deep in the bilayer.

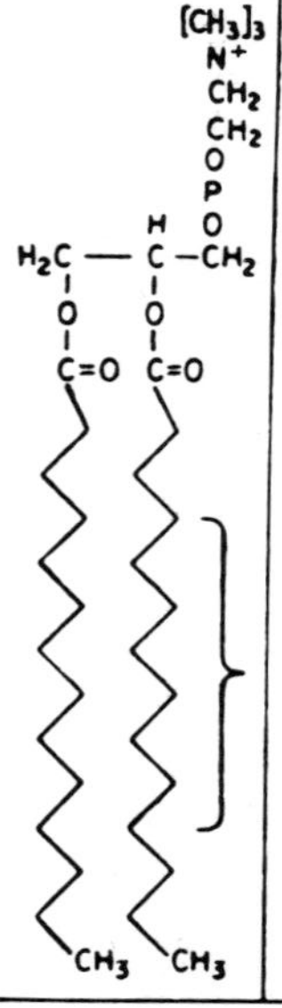

(a)

Phosphatidylcholine	Tm^{3+} chemical shift (ppm)	Mn^{2+} T_1 RE (s^{-1})	Lipid spin label T_1 RE (s^{-1})
[CH$_3$]$_3$ N$^+$	3.2	14	0.1
CH$_2$	0.1	12	0.2
CH$_2$	0.2	8	0.0
O P O	0.2	3	0.2
H$_2$C—C—O—C=O / O=C—O—CH$_2$			0.2 / 1.1
(carbons 5–10)	0.2	0.9	0.8
11 / 12	0.0	0.4	0.6
13	0.1	0.4	0.6
14 CH$_3$ CH$_3$	≡0.0	0.3	0.8

(b)

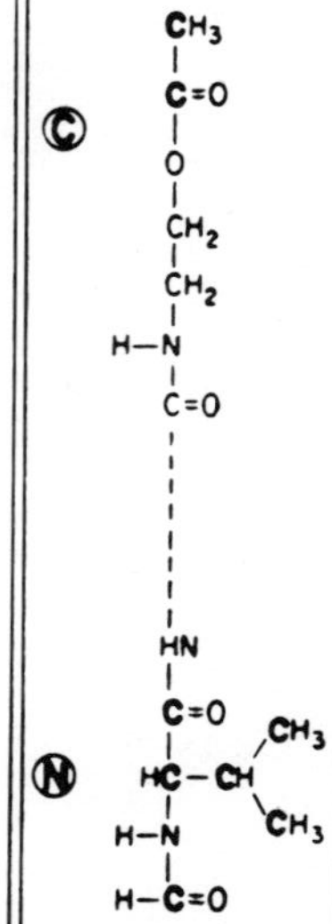

[N, C-^{13}C] gramicidin	Tm^{3+} chemical shift (ppm)	Mn^{2+} T_1 RE (s^{-1})	Lipid spin label T_1 RE (s^{-1})
CH$_3$—C=O—O—CH$_2$—CH$_2$—N(H)—C=O (C terminus)	1.8 / 2.0	1.9	0.0
HN—C=O—HC—CH(CH$_3$)(CH$_3$) (N terminus)	0.5	0.7	1.9
H—N—H—C=O	0.4	0.4	0.8

Fig. 7. Summary of NMR probe induced changes in chemical shift and T_1 rate enhancements of phosphatidylcholine and gramicidin resonances. (a) The dimyristoyl phosphatidylcholine molecule is depicted with its fatty acid chain in the extended configuration for simplicity. In actuality, the NMR measurements were carried out above the phase transition temperature, so that the chains are disordered and highly mobile. (b) The linear sequence of [N,C^{13}-C] gramicidin is indicated with the C terminus at the top and the N terminus at the bottom. The chemical structures of the termini containing the ^{13}C labels are shown with the intervening 15 amino acids omitted. These data strongly indicate that gramicidin molecules in phospholipid bilayers form dimer channels oriented with their C termini close to the surface and their N termini buried deep within the hydrophobic regions of the bilayer (Weinstein, (1980)).

C. Permeability Studies

The nondestructive nature of NMR experiments and the kinetic information that can be extracted from line shape analysis and nuclear relaxation rates make NMR spectroscopy an important and useful tool in permeability studies (Degani, 1978b; Degani and Bar-On, 1981).

There are two types of transfer processes that can be monitored by NMR. The first type involves flux measurements in which the flow of material can be followed by observing time-dependent changes in the integrated intensity of the magnetic resonance signal due to nuclei in the permeating molecules, or in some other component of the system affected by the permeation. This method is applicable to relatively slow transport processes with half-lives of more than a few seconds. The second type corresponds to the exchange of molecules across a membrane barrier. The theoretical basis of such studies is the same as for chemical exchange (Section III,B). The dynamic information is manifested by changes in the chemical shift, line shape, and nuclear relaxation times T_1 and T_2 of nuclei belonging to each of the exchanging molecules. These studies enable one to follow fast permeation kinetics, with half-lives ranging from seconds to tenths of milliseconds. It should be noted that the saturation transfer method (Forsen and Hoffman, 1963) is suitable for studying exchange processes with intermediate rates (on the order of the T_1 relaxation rate of the observed nucleus). This method, however, has not yet been used for permeability measurements.

A necessary requirement for determining membrane permeability with NMR spectroscopy is the possibility of distinguishing between the signals on each side of the membrane. This separation can be induced by adding nonpenetrating paramagnetic reagents (as discussed later) or by employing internal field gradients (Brindle et al., 1979) or externally induced field gradients (Andrasko, 1976; Cooper et al., 1974).

The first NMR demonstration of an ionophore-mediated ion translocation across the membranes of unilamellar lipid vesicles was reported by Fernandez et al. (1973). Their approach was based on observing transport-induced changes in the NMR signal of the lipid nuclei on each side of the membrane. On adding the ionophore X-537A the inner methylcholine signal collapsed with the outer signal, originally shifted by extravesicular Pr^{3+} ions. Hunt (1975) suggested a determination of Pr^{3+} transport rates mediated by the ionophores X-537A and A23187, by following the time evolution of the broadening of the inner methylcholine line. It was later reported that such rates of ion transport could not be determined directly from the time-dependent broadenings or shifts, owing to the uneven distribution of the Pr^{3+} ions inside the very small vesicles used in such studies (Hunt *et al.*, 1978; Ting *et al.*, 1981). The kinetics of the transport of ions that act as shift reagents of membranal NMR signals, such as Pr^{3+}, can therefore be determined by full line shape analysis of the relevant signals (Ting *et al.*, 1981). Another possibility that bypasses the need for a detailed line shape analysis is the use of a calibration curve (Degani, 1978a; Hunt *et al.*, 1978). This curve is obtained by monitoring the inner choline signal in vesicular suspensions containing varying intravesicular concentrations of the paramagnetic ion. It was pointed out that a unique situation exists when the transport of strong relaxing paramagnetic ions is followed in small vesicles (Degani and Lenkinski, 1980; Degani *et al.*, 1981). The transport of only one paramagnetic ion into the intravesicular space produces there a concentration to such an extent that the signal due to the inner methylcholine protons is substantially broadened. The observed signal is due to vesicles that do not contain paramagnetic ions. As ion transport proceeds, this signal decreases in intensity but does not broaden (Fig. 8). Degani *et al.* (1981) have shown that for Mn^{2+} transport the kinetics can be determined directly from the initial time dependence of the intensity of the signal. This method has a general and important advantage in that it measures the transport of a single ion without the complications associated with back-diffusion of the ionophore.

In addition to quantitative rate data the kind of mechanistic information that can be obtained from NMR transport measurements is demonstrated in the study on X-537A-mediated Mn^{2+} transport across dipalmitoyl phosphatidylcholine vesicles (Degani *et al.*, 1981). By monitoring the depen-

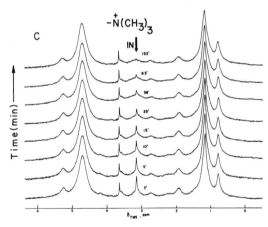

Fig. 8 The X-537 A mediated Mn^{2+} transport: 270-MHz ^{1}H NMR spectra of egg phosphatidylcholine vesicles prepared in ^2H_2O containing 6.7 mM Tris, in the presence of 30μM X-537A and 1 mM MnCl$_2$ added after sonication at pH = 6.8 and 28°C. Chemical shifts were referenced to an external tetramethylsilane (Degani, 1978).

dence of the rates k' on the X-537A concentration over a broad range it was concluded that manganese is transported to the vesicles via both 1 : 1 and 2 : 1 ionophore-Mn^{2+} complexes. At low X-537A/vesicle concentration ratios (<1) the 1 : 1 complex predominates, whereas at high ratios (>2) the 2 : 1 complex predominates. From the dependence of k' on the manganese concentration it was concluded that equilibration of the ionophore X-537A between the vesicles is much faster than the transport rate through the vesicular membrane. Temperature dependence studies indicated that ionophore-mediated ion transport was very sensitive to the lipid liquid crystalline–gel phase transition (Fig. 9). On passing from the liquid crystalline to the gel phase, a reduction of about two orders of magnitude in the transport rate and a large increase in the transport activation energy from 9 kcal/mol to approximately 50 kcal/mol are observed.

The study of ionophore-mediated fluxes of paramagnetic ions by monitoring the proton signal of the methylcholine groups in the membrane was

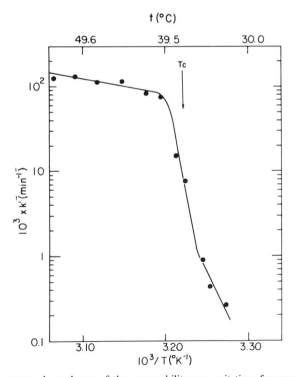

Fig. 9 Temperature dependence of the permeability per unit time for manganese transport in sonicated dipalmitoyl phosphatidylcholine vesicles. The vesicles suspension in ^2H_2O contained: 6.7 mM Tris chloride, p^{2}H = 6.8, 71μM X-537A and 1 mM MnCl$_2$. (Degani *et al.*, 1981).

extended to ^{31}P-NMR measurements of the phosphate moiety. These measurements were aimed at determining the effect of antibiotics such as nystatin (Pierce *et al.*, 1978) and etheromycin (Donis *et al.*, 1981) on Pr^{3+} permeability, and X-537A on Co^{2+} permeability (Degani, 1978b), in lipid

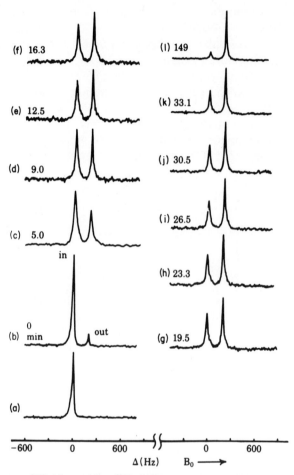

Fig. 10 Transport of Na^+ in vesicles. ^{23}Na NMR spectra (132.3 MHz) of a dispersion of large unilamellar vesicles in ^2H_2O containing gramicidin A. (a) In the NMR tube, the final concentrations were: egg phosphatidylcholine 6.0 mM, $NaCl_{in}$ 60 mM; $NaCl_{out}$ 0.27 mM; and $LiCl_{out}$, 35 mM. (b) Concentrations were as in (a) except that the outside aqueous space was made 6.3 mM in $[HN(CH_2CH_2OH)_3]_3$ $Dy[N(CH_2CO_2)_3]_2$. For (a) and (b), 512 FID's were accumulated in 208 sec. (c-1) spectra are labeled with the minutes elapsed after the solution was made 0.16 M in gramicidin. The times recorded are those of the midpoints of the data accumulation periods, which were 156 FID in 64 s (c) 128 FID in 52 s (d-j), and 512 FID in 208 s (k and l). The temperature was ≈297K (Pike M. M. *et al.*, 1982)].

vesicles. The analysis of the data was essentially the same as for the protons.

Recently sodium ion fluxes were measured directly with ^{23}Na NMR (Pike *et al.*, 1982). Gramicidin-induced sodium diffusion into and out of large unilamellar vesicles was followed by monitoring changes with time in the integrated intensity of the outer and inner signals that were shifted from one another by dysprosium nitrolotriacetate ion, Dy[N-$(CH_2COO)_3]_2^{3-}$ (Fig. 10).

Nuclear magnetic resonance measurements of exchange processes across membranes were initially used to follow water diffusion (Conlon

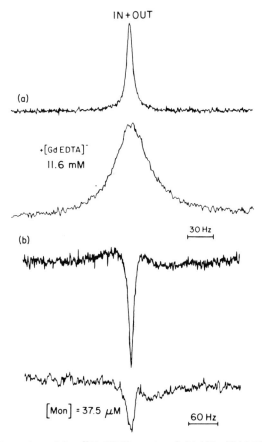

Fig. 11 Na$^+$ exchange in vesicles. ^{23}Na NMR spectra of: (a) 150 mM NaCl in phosphatidyl-choline vesicles suspension, p^{2}H 9.2 at 28°C, and after addition of 11.6 mM Gd(EDTA)$^-$ to the extravesicular medium. (b) The same as (a) with Gd(EDTA)$^-$ using 180°-τ-90° pulse sequence in order to separate the signal due to the ions entrapped in the inner vesicular medium, without and with added monensin (Degani and Elgavish, 1978).

and Outhred, 1972). Degani and Elgavish (1978) extended this method to ion transport studies by monitoring the metal ion NMR directly (e.g., ^{7}Li and ^{23}Na). In this study it was realized that owing to electrostatic repulsion the common shift and relaxation reagents, which are positively charged paramagnetic ions, have only minor effects on the NMR characteristics of nuclei belonging to positively charged ions. On the other hand, a negatively charged paramagnetic complex such as the ethylenediamine tetraacetate (EDTA) complex of Gd^{3+} interacts with the positive ions and markedly enhances their T_1 and T_2 nuclear relaxation rates. By using a pulse sequence of 180°-τ-90° with τ equal to the time needed to nullify the enhanced relaxation, it was possible to distinguish between the ions on both sides of the membrane (Fig. 11). This method was employed in studying monensin-mediated sodium and lithium transport across phosphatidylcholine vesicles.

A combination of these methods was recently employed in studying X-537A-mediated exchange of lanthanide ions between hydrated phospholipid micelles in benzene (Chen and Springer, 1981). The distinct ^{31}P signals from phospholipid molecules in micelles containing no paramagnetic ions and in micelles containing a single Pr^{3+} or Eu^{3+} ion were exchange-broadened in the presence of the ionophore. The average preexchange lifetimes as a function of antibiotic concentration were determined by line shape analysis. The results seem to indicate that the mechanism of the ionophore-mediated ion transfer in inverted micelles differs from that in lipid vesicles.

The time has come to proceed to characterization of the ionophores and their activities under physiological conditions using the recent intensive developments in *in vivo* NMR studies.

References

Andrasko, J. (1976). *Biochim. Biophys. Acta* **428,** 304–311.

Anteunis, M. J. O. (1981). *Bull. Soc. Chim. Belg.* **90,** 449–470.

Anteunis, M. J. O. (1982). In "Polyether Antibiotics" (J. W. Westley, ed.), Vol. II, chapter 15. Dekker, New York.

Berger, J., Rachlin, A. I., Scott, W. E., Sternbach, L. H., and Goldberg, M. W. (1951). *J. Am. Chem. Soc.* **73,** 5295–5298.

Borremans, F., and Anteunis, M. J. O. (1981). *Bull. Soc. Chim. Belg.* **90,** 1045–1053.

Briggs, R. W., and Hinton, J. F. (1978a). *J. Magn. Reson.* **32,** 155–160.

Briggs, R. W., and Hinton, J. F. (1978b). *Biochemistry* **17,** 5576–5582.

Briggs, R. W., and Hinton, J. F. (1979). *J. Magn. Reson.* **33,** 363–377.

Briggs, R. W., Etzkorn, F. A., and Hinton, J. F. (1980). *J. Magn. Reson.* **37,** 523–528.

Brindle, K. M., Brown, F. F., Campbell, I. D., Grathwohl, C., and Kuchel, P. W. (1979). *Biochem. J.* **180,** 37–50.

Brockmann, H., and Schmidt-Kastner, G. (1955). *Chem. Ber.* **88**, 57–61.

Buckley, P. D., Jolley, K. W., and Pinder, D. N. (1975). *Prog. NMR Spectroscopy* **10**, 1–26.

Bystrov, V. F. *et al.* (1972). *FEBS Lett.* **21**, 34–38.

Bystrov, V. F., Gavrilov, Y. D., Ivanov, V. T., and Ovchinnikov, Y. A. (1977). *Eur. J. Biochem.* **78**, 63–82.

Chan, S. I., Feigenson, G. W., and Seiter, C. H. A. (1971). *Nature* **231**, 110–112.

Chen, S. T., and Springer, C. S. Jr. (1978). *Bioinorg. Chem.* **9**, 101–122.

Chen, S. T., and Springer, C. S. Jr. (1981). *Biophys. Chem.* **14**, 375–388.

Civan, M. H., and Shporer, M. (1978). In "Biological Magnetic Resonance" (L. J. Berliner and J. Reuben, eds.) Vol. 1, pp. 1–32. Plenum Press, New York.

Conlon, T., and Outhred, R. (1972). *Biochim. Biophys. Acta* **288**, 354–361.

Cooper, R. L., Chang, D. B., Young, A. C., Martin, C. J., and Ancker-Johanson, B. (1974). *Biophys. J.* **14**, 161–177.

Cornell, B. A., Keniry, M., Hiller, R. G., and Smith, R. (1980). *FEBS Lett.* **115**, 134–138.

Cornelis, A., and Laszlo, P. (1979). *Biochemistry* **10**, 2004–2007.

Davis, D. G. (1975). *Biochem. Biophys. Res. Commun.* **63**, 786–791.

Davis, D. G., and Tosteson, D. C. (1975). *Biochemistry* **14**, 3962–3969.

Degani, H. (1977). *Biophys. Chem.* **6**, 345–349.

Degani, H. (1978a). *Biochem. Biophys. Acta* **508**, 364–369.

Degani, H. (1978b). In "NMR Spectroscopy in Molecular Biology" (B. Pullman, ed.), pp. 393–403. Reidel, Dordrcht, Holland.

Degani, H., and Bar-On, Z. (1981). *Period. Diol.* **83**, 61–68.

Degani, H., and Elgavish, G. A. (1978). *FEBS Lett.* **90**, 357–360.

Degani, H., and Friedman, H. L. (1974). *Biochemistry* **13**, 5022–5032.

Degani, H., and Friedman, H. L. (1975). *Biochemistry* **14**, 3755–3761.

Degani, H., Hamilton, R. M. D., and Friedman, H. L. (1976). *Biophys. Chem.* **4**, 363–366.

Degani, H. and Lenkinski, R. E. (1980). *Biochemistry* **19**, 3430–3434.

Degani, H., Sanford, S., and McLaughlin, A. C. (1981). *Biochem. Biophys. Acta* **646**, 320–328.

Delville, A., Detellier, C., and Laszlo, P. (1979). *J. Magn. Reson.* **34**, 301–315.

Donis, J., Grandjean, J., Grosjcan, A., and Laszlo, P. (1981). *Biochim. Biophys. Res. Commun.* **102**, 690–696.

Duax, W. L., Hauptman, H., Weeks, C. M., and Norton, D. A. (1972). *Science* **176**, 911–914.

Feigenson, G. W., and Meers, P. R. (1980). *Nature* **283**, 313–314.

Feigenson, G. W., Meers, P. R., and Kingsley, P. B. (1977). *Biochim. Biophys. Acta* **471**, 487–491.

Fernandez, M. S., Celis, H., and Montal, M. (1973). *Biochim. Biophys. Acta* **343**, 600–605.

Finer, E. G., Hauser, H., and Chapman, D. (1969). *Chem. Phys. Lipids* **3**, 386–392.

Forsen, S., and Hoffman, R. A. (1963). *J. Chem. Phys.* **39**, 2892–2901.

Forsen, S., and Lindman, B. (1981). In "Methods of Biochemical Analysis" (D. Glick, ed.), Vol. 27, pp. 289–472. Wiley, New York.

Funck, Th., Eggers, F., and Grell, E. (1972). *Chimia* **25**, 637–641.

Gambale, F., Gliozzi, A., and Robello, M. (1973). *Biochim. Biophys. Acta* **330**, 325–334.

Glickson, J. D., Gordon, S. L., Pitner, T. P., Agresti, D. G., and Walter, R. (1976). *Biochemistry* **15**, 5721–5729.

Gordon, L. G. M., and Haydon, D. A. (1972). *Biochim. Biophys. Acta* **225**, 1014–1018.

Grandjean, J., and Laszlo, P. (1979). *Angew. Chem. Int. Engl. Ed.* **18**, 153–154.

Grell, E., and Funck, T. (1973). *Eur. J. Biochem.* **34**, 415–424.

Hauser, H., Finer, E. G., and Chapman, D. (1970). *J. Mol. Biol.* **53**, 419–433.

Haynes, D. H. (1972). *FEBS Lett.* **20**, 221–224.

Haynes, D. H., Kowalsky, A., and Pressman, B. C. (1969). *J. Biol. Chem.* **244**, 502–505.

Hladky, S. B., and Haydon, D. A. (1970). *Nature* **225**, 451–453.

Hsu, M., and Chan, S. I. (1973). *Biochemistry* **12**, 3872–3876.

Hunt, G. R. A. (1975). *FEBS Lett.* **58**, 194–196.

Hunt, G. R. A., Tipping, L. R. H., and Belmont, M. R. (1978). *Biophys. Chem.* **8**, 341–355.

Ivanov, V. T. *et al.* (1969). *Biochem. Biophys. Res. Commun.* **34**, 803–811.

Ivanov, V. T. *et al.* (1971). *Khimiya Prirodnykh Soedinenii* **7**, 217–236 (Russ. 221–245).

Kaplan, J. I., and Fraenkel, G. (1980). "NMR of Chemically Exchanging Systems". Academic Press, New York.

Kleier, D. A., and Binsch, G. (1970). *J. Magn. Resonan.* **3**, 146–160.

Kopple, K. D., Ohnishi, M., and Go, A. (1969). *J. Am. Chem. Soc.* **91**, 4264–4272.

Krishnan, C. V., Friedman, H. L., and Springer, C. S., Jr. (1978). *Biophys. Chem.* **9**, 23–35.

Laszlo, P. (1978). *Angew. Chem. Int. Engl. Ed.* **17**, 254–266.

Lau, A. L. Y., and Chan, S. I. (1974). *Biochemistry* **13**, 4942–4948.

Lau, A. L. Y., and Chan, S. I. (1975). *Proc. Natl. Acad. Sci. USA* **72**, 2170–2174.

Lenkinski, R. E., and Reuben, J. (1976). *J. Magn. Reson.* **21**, 47–56.

Loewenstein, A., and Connor, T. M. (1962). *Ber. Bunsenges Physik. Chem.* **67**, 280–295.

McLaughlin, S., and Eisenberg, M. (1975). *Ann. Rev. Biophys. Bioeng.* **4**, 335–366.

Mayers, D. F., and Urry, D. W. (1972). *J. Am. Chem. Soc.* **94**, 77–81.

Moore, C. and Pressman, B. C. (1964). *Biochem. Biophys. Res. Commun.* **15**, 562–567.

Neupert-Laves, K., and Dobler, M. (1975). *Helv. Chim. Acta* **58**, 432–442.

Ohanishi, M., and Urry, D. W. (1969). *Biochem. Biophys. Res. Commun.* **36**, 194–202.

Ohanishi, M., and Urry, D. W. (1970). *Science* **168**, 1091–1092.

Ohanishi, M., Fedarko, M. C., Baldeschwieler, J. D., and Johanson, L. F. (1972). *Biochem. Biophys. Res. Commun.* **46**, 312–320.

Oldfield, E. *et al.* (1978). *Proc. Natl. Acad. Sci. USA* **75**, 4657–4660.

Ovchinnikov, Yu. A., Ivanov, V. T., and Shkrob, A. M. (1974). In "Membrane Active Complexones", Vol. 12. Elsevier, New York.

Patel, D. J. (1973). *Biochemistry* **12**, 496–501.

Patel, D. J., and Shen, C. (1976). *Proc. Natl. Acad. Sci. USA* **73**, 1786–1790.

Patel, D. J., and Tonelli, A. E. (1973). *Biochemistry* **12**, 486–496.

Pierce, H. D., Unrau, A. M., and Oehlschlager, A. C. (1978). *Can. J. Biochem.* **56**, 801–807.

Pike, M. M., Simon, S. R., Balschi, J. A., and Spriger, C. S., Jr. (1982). *Proc. Natl. Acad. Sci. USA* **79**, 810–814.

Pinkerton, M., Steinrauf, L. K., and Dawkis, P. (1969). *Biochem. Biophys. Res. Commun.* **35**, 512–518.

Pressman, B. C. (1976). *Ann. Rev. Biochem.* **45**, 501–530.

Pressman, B. C., Harris, E. J., Jagger, W. S., and Johanson, J. H. (1967). *Proc. Natl. Acad. Sci. USA* **58**, 1949–1959.

Prestegard, H., and Chan, S. I. (1969). *Biochemistry* **8**, 3921–3927.

Prestegard, J. H., and Chan, S. I. (1970). *J. Am. Chem. Soc.* **92**, 4440–4446.

Reed, P. W., and Lardy, H. A. (1972). *J. Biol. Chem.* **21**, 6970–6977.

Schori, E., Jagur-Grodzinski, J., Luz, A., and Shporer, M. (1971). *J. Am. Chem. Soc.* **93**, 7133–7138.

Shen, C., and Patel, D. J. (1976). *Proc. Natl. Acad. Sci. USA* **73**, 4277–4281.

Shen, C., and Patel, D. J. (1977). *Proc. Natl. Acad. Sci. USA* **74**, 4734–4738.

Shporer, M., Zemel, H., and Luz, Z. (1974). *FEBS Lett.* **40**, 357–360.

Smith, G. D. *et al.* (1975). *J. Am. Chem. Soc.* **97**, 7242–7247.

Stark, G., Ketterer, B., Benz, R., and Läuger, P. (1971). *Biophys. J.* **11**, 981–994.

Swift, T. J., and Connick, R. E. (1962). *J. Chem. Phys.* **37**, 307–320.

Ting, D. Z., Hagan, P. S., Chan, S. I., Doll, J. D., and Springer, C. S., Jr. (1981). *Biophys. J.* **34**, 189–216.

Weinstein, S., Wallace, B. A., Blout, E. R., Morrow, J. S., and Veatch, W. (1979). *Proc. Natl. Acad. Sci. USA* **76**, 4230–4234.

Weinstein, S., Wallace, B. A., Morrow, J. S., and Veatch, W. R. (1980). *J. Mol. Biol.* **143**, 1–19.

Westley, J. W., Evans, R. H., Jr., Williams, T., and Stempel, A. (1970). *Chem. Commun.* 71–72.

Woessner, D. E. (1961). *J. Chem. Phys.* **35**, 41–48.

Index

289